Climate Future

Climate Future

Averting and Adapting to Climate Change

Robert S. Pindyck

OXFORD
UNIVERSITY PRESS

Oxford University Press is a department of the University of Oxford. It furthers the University's objective of excellence in research, scholarship, and education by publishing worldwide. Oxford is a registered trade mark of Oxford University Press in the UK and certain other countries.

Published in the United States of America by Oxford University Press
198 Madison Avenue, New York, NY 10016, United States of America.

Library of Congress Cataloging-in-Publication Data

Names: Pindyck, Robert S., author.
Title: Climate future : averting and adapting to climate change /
Robert S. Pindyck.
Description: New York, NY : Oxford University Press, [2022] |
Includes bibliographical references and index.
Identifiers: LCCN 2021058986 (print) | LCCN 2021058987 (ebook) |
ISBN 9780197647349 (hardback) | ISBN 9780197647363 (epub)
Subjects: LCSH: Climatic changes. | Climatic changes—Government policy. |
Climate change mitigation. | Greenhouse gases—Environmental aspects. |
Carbon dioxide—Environmental aspects. | Greenhouse gas mitigation.
Classification: LCC QC903 .P555 2022 (print) | LCC QC903 (ebook) |
DDC 363.738/74561—dc23/eng20220317
LC record available at https://lccn.loc.gov/2021058986
LC ebook record available at https://lccn.loc.gov/2021058987

Printed in Canada by Marquis Book Printing

To my grandchildren,
Noa, Alma, Revi, Lev, and those to come.

Contents

Preface

This book grew out of research on climate change (and environmental economics more generally) that I have been engaged in over the past decade. The primary focus of this research has been on the nature and implications of uncertainty. How much do we know or not know about climate change itself, and about its impact on the economy and on society in general? Has the uncertainty over climate change and its impact been increasing or decreasing over the past decade or two, and how is it likely to change in the future? And what does this uncertainty imply for climate policy? Does it mean that we should avoid taking drastic action now, and instead wait until we learn more? Or does it mean the opposite; that we should act now, as a way of buying insurance against the possibility of a very bad climate outcome?

This book is also a response to the fact that many of the books, articles, and press reports that we read make it seem that we know a lot more about climate change and its impact than is actually the case. Commentators and politicians often make statements of the sort that if we don't sharply reduce CO_2 emissions, the following things will happen, as though we actually knew what will happen. Rarely do we read or hear that those things *might happen*; instead we're told they *will happen*. Naturally, we humans prefer certainty to uncertainty, and feel uncomfortable when we don't know what lies ahead. Most people prefer to hear or read statements of the sort "By 2050 the temperature will rise by 3°C and sea levels will rise by 5 meters," as opposed to "there is a 30-percent chance that temperature will rise by 3°C or more, and a 70-percent chance it will rise by less than 3°C." But as distressing as it might be, the simple fact is that the "climate outcome," by which I mean the extent of climate change and its impact on the economy and society more generally, is far more uncertain than most people think.

I feel that it is important for people to better understand the extent and nature of the uncertainty, and why it is that while there are certain things that we do know about climate change, there are other things that we don't know, and may not know for a long time, if ever. This book provides what I hope is an accessible explanation of what we know and don't know, and the nature and implications of the uncertainty we face.

This book is also the result of a related area of research, and that is my work on the economic and policy implications of potential catastrophes. Our society faces a variety of potential global catastrophic events—nuclear or bio-terrorism and major pandemics are examples. But another example is a climate catastrophe, by which I mean a climate outcome involving severe social disruption and in which the economy experiences a major contraction. In fact, as I argue in this book, the possibility of a climate catastrophe is (or should be) the main driver of climate policy.

What This Book Is About

Current debates over climate change policy are focused almost entirely on how to reduce emissions. Reducing emissions—via a carbon tax, emission quotas, adoption of "green" energy technologies, or other means—is an important goal, and should continue to be a fundamental part of climate policy. It should also be a subject of environmental policy research; we need to learn more about *how* to reduce emissions, and what are the advantages and disadvantages of alternative approaches to reducing emissions.

Reducing emissions is clearly something we should do. But we also need to answer the following question: *What will we do?* Yes, in all likelihood we will reduce emissions of CO_2 and other greenhouse gases, but by how much and how fast? Enough to prevent a temperature increase greater than 1.5 or 2.0°C by the end of the century? What if, despite our best efforts, we are simply unable to reduce emissions enough to prevent a temperature increase of 2°C or more? What then? Do we just wave our hands and say "Too bad"? And if today we think that there is a strong possibility that (despite our best efforts) we are likely to experience a temperature increase greater than 2°C, what should we do?

I argue in this book that given the political and economic realities, it is extremely unlikely that the world will even come close to meeting current targets for CO_2 emission reductions. Some countries (e.g., the U.S. and Europe) may meet their targets, but other countries (China, India, Indonesia, Russia, . . .) will not, and may not even set targets. In fact, even the most optimistic projections of CO_2 emission reductions imply a substantial buildup of CO_2 in the atmosphere, and as a result a gradual increase in temperatures worldwide.

I explain some of the basics of climate change in this book, with an emphasis on what we know and don't know about the extent of climate change that might occur, and the impact of climate change on the economy and on society more generally. Despite decades of research, there is still a great deal that we don't know about climate change, and perhaps most important, about the impact that higher temperatures and rising sea levels might have. Put simply, whatever climate policies are adopted, there will be a great deal of uncertainty over what will happen as a result. I explain why there is so much uncertainty, and what it implies for the design of policy.

If the world succeeds in reducing CO_2 emissions substantially, what will happen? As I just said, we don't know exactly, and can only project a range of possible outcomes. Nonetheless, I show that for any realistic scenario for CO_2 emission reductions, there is a strong likelihood of a global mean temperature increase over the next 50 years that could turn out to be 3°C or even higher. Higher temperatures could lead to rising sea levels, greater variability of weather, more intense storms, and other forms of climate change. What will be the impact of those changes? How will it affect economic output and other measures of social welfare, such as mortality and morbidity? Again, we don't know. But the fact that we don't know doesn't mean we should be complacent. The outcome could be catastrophic, especially if society is unprepared for it.

What does this reality imply for climate policy? I argue that partly because there is so much uncertainty, we need to do more to reduce emissions. But reducing emissions is not enough; to insure against a catastrophic climate change outcome we need to invest now in *adaptation*. Adaptation can have many forms—developing new hybrid crops, adopting policies to discourage building in flood-prone or wildfire-prone areas, building sea walls and dikes, and forms of geoengineering are examples. Developing new ways to abate emissions remains important, but climate change research, and climate change policy, should put more emphasis on adaptation than it has in the past.

Acknowledgments

I have been fortunate to have a good number of colleagues and friends who provided ideas, suggestions, and comments as this book took shape. But I am especially indebted to the late Martin Weitzman, without whom this book would never have happened. Marty was one of the world's leading environmental economists, and his seminal papers and books became a bedrock of environmental economics in general, and the economics of climate change in particular. Although he was probably best known for his work in environmental economics, he made important contributions to other areas of economics as well. When Marty passed away on August 27, 2019, the world lost one of its most original and productive thinkers.

I said that without Marty, this book would never of happened. I had originally planned to write a policy-oriented paper on climate change, and to that end I wrote a detailed outline which I shared with numerous people. I received lots of useful comments and suggestions, for which I am grateful, but Marty guided me in a direction that others didn't. He told me that what I needed to say wouldn't fit in a paper, even a very long one—it required a book. No, forget the paper, it has to be a book, and he gave me some excellent ideas as to what the book should look like. I could have argued with Marty about this (as I argued with him about many ideas in economics, without ever getting him to change his mind), but I knew that he was right, so I better get busy writing.

Many other people also contributed in various ways to this book. Gilbert Metcalf provided very detailed comments on an earlier draft of the manuscript, and pointed out a number of errors that I needed to correct. Dick Schmalensee went through the manuscript line by line, found a number of errors, and made many helpful suggestions. Sergio Franklin, with whom I had done research earlier on the social cost of deforestation, provided extensive comments and guidance for the section on forests and their impact on net CO_2 emissions. I received detailed comments and had extensive discussions with John Deutch, which led me to revise several parts of the book, and especially the section on nuclear power. Sergio Vergalli also went through the manuscript line by line and provided detailed comments. Alan Olmstead provided suggestions regarding adaptation in agriculture, as well as Figure 7.1 on wheat production in the U.S. during the 1850s. Christian Gollier, Geoff Heal, Matt Kotchen, Chuck Manski, Richard Newell, and Cass Sunstein all read through the draft and provided numerous suggestions for ways to improve it. I also received very useful comments and suggestions from Edward Dlugokencky, Stephanie Dutkiewicz, Chris Forest, Kenneth Gillingham, Henry Jacoby, Chris Knittel, Bob Litterman, John Lynch, Sergey Paltsev, Ron Prinn, Mala Radhakrishnan, John Reilly, Andrei Sokolov, Susan Solomon, Rob Stavins, Jim Stock, John Sterman, Richard Tol, and Gernot Wagner.

I also received valuable help from MIT's Joint Program on the Science and Policy of Global Change. The Joint Program has developed an Economic Projection and Policy Analysis (EPPA) model that simulates how human activities affect greenhouse gas emissions (and other air and water pollutants), which are inputs into an Earth System Model that simulates the resulting changes in the physical processes occurring in the atmosphere. Sergey Paltsev and Andrei Sokolov ran the MIT Earth System Model (MESM) several times to help explore the possible impacts of large pulses of CO_2 emissions.

My thanks to Bob Litterman, who provided financial support in the form of a fund used to hire research assistants. Jack Barotta worked as an RA during the early stages of this project, and helped gather information on solar geoengineering, as well as other forms of adaptation. I am especially indebted to Miray Omurtak, who has provided invaluable research assistance over the past two years. Miray worked on all aspects of this project, from writing MATLAB programs to simulate temperature trajectories, to investigating alternative approaches to modeling CO_2 and methane emissions, to creating a database of emissions and concentrations across countries, to studying and modeling the advantages, disadvantages, and costs of alternative approaches to adaptation. It's hard for me to imagine how this book could have been completed without her help.

And of course at the end of the day the usual disclaimer applies: I, and only I, am to blame for all of the errors, large and small, that might appear in this book.

1

Introduction

Most books, articles, and discussions about climate change focus on two very important questions. First, if the world continues to emit growing amounts of greenhouse gases (GHGs), what will happen to the climate over the coming decades? By how much will temperatures increase? What will warming do to sea levels, the severity and frequency of storms and hurricanes, the extent of droughts, and other aspects of climate? And, perhaps most important, what will be the economic and social damage resulting from these changes?

Second, what should be done to avert climate change? In particular, by how much and how rapidly should GHG emissions be reduced, and what policy tools should be used to achieve those emission reductions? Is a carbon tax the best policy tool, and if so how large should the tax be?

But there are two additional questions that are equally important. First, while we might agree on what *should* be done, we need to ask what *will be done* to avert or reduce the extent of climate change. Even if we are optimistic about the likelihood of countries agreeing to major reductions in their GHG emissions, what kinds of emission scenarios can we realistically expect to see? Is it reasonable to think that worldwide emissions will fall drastically and rapidly enough during the next few decades to prevent severe climate change?

Second, suppose we conclude that it is *not* realistic to expect global GHG emissions to fall sufficiently and quickly enough, so that despite our best efforts we (or our children and grandchildren) are likely to experience higher temperatures and rising sea levels. Then what should we do in response? Should we take actions now to avert or reduce the impact of climate change that is likely to result given realistic emission scenarios, and if so, what kinds of actions?

This second set of questions is a major focus of this book. I will explain that there is a great deal of uncertainty over what might happen, and we might be lucky and end up with only a mild degree of climate change. But counting on good luck does not make for smart policy. The fact is that given the economic and political realities, it is simply *not realistic* to expect the kinds of GHG emission reductions needed to avert a substantial amount of global warming, and as a result, we *should take actions now to reduce the possible impacts of*

Climate Future: Averting and Adapting to Climate Change. Robert S. Pindyck, Oxford University Press.
© Oxford University Press 2022. DOI: 10.1093/oso/9780197647349.003.0001

that warming. The actions I am referring to involve various forms *adaptation.* How do I reach those conclusions, and what forms of adaption do I have in mind? Read on to find out.

1.1 Averting and Adapting: The Basic Argument

The basic argument of this book is fairly simple, and can be summarized in terms of the following six points. Most readers will (I hope) readily agree with the first three points, but I expect fewer will agree with the last three—at least until they have finished reading the book. The argument goes as follows:

(1) GHG Emissions and Climate Change

First, there is little disagreement that the world is continuing to emit large amounts of greenhouse gases (GHGs), mostly in the form of carbon dioxide (CO_2), but also methane and other gases. These GHG emissions have been growing steadily over the past century, and emissions of CO_2 in particular will remain in the atmosphere for centuries to come. As a result, the atmospheric concentrations of CO_2 and other GHGs have also been growing. There is also little disagreement that as these GHG emissions accumulate in the atmosphere, they will eventually lead to *climate change*: a general warming of the planet, which in turn can result in rising sea levels, stronger and more frequent storms and hurricanes, more severe droughts, and more extreme weather conditions in many parts of the world.

In fact, because of GHG emissions over the past century, we are already witnessing an increase in the global mean temperature, as shown in Figure 1.1. Since 1960, the temperature has increased by something close to 1.0°C. ("C" stands for Celsius, and 1.0°C is equivalent to 1.8 degrees Fahrenheit (1.8°F).) Furthermore, as the figure makes clear, most of this increase in temperature occurred after 1980, i.e., in just the last 40 years. The world is warming, and the rate of warming seems to be accelerating. And although we can't be sure, these temperature increases may be at least partly responsible for some of the past decade's more extreme weather.

How much more climate change should we expect in the future, and how soon will it occur? We don't know. It depends in part on the climate system, which we don't fully understand. And of course it depends on the world's GHG emissions over the coming decades, which in turn depends on what kinds of policies are adopted to reduce emissions. In the near term—the next decade or two—world emissions are likely to keep increasing, despite ongoing efforts in the U.S., Europe, Japan, and other countries to reduce them.

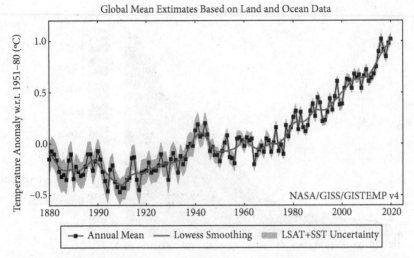

Fig. 1.1 Global Change in Temperature from 1880 to 2020. Jagged line is global annual mean, and smoothed line shows five-year averages. Note that most of the temperature increase occurred after 1970.
Source: NASA, GISS Surface Temperature Analysis.

Why should we expect world GHG emissions to keep growing, at least over the coming decade? After all, the U.S. and Europe have already made progress in reducing emissions, and are likely to make more progress. The problem can be seen in Figure 1.2, which shows CO_2 emissions by region. Observe the rapid growth of emissions from China, India, and other Asian countries such as Malaysia, Indonesia, Thailand, and Vietnam. Prior to 1980, these countries contributed a relatively small amount to worldwide CO_2 emissions, largely because their economies were less developed. But as their economies picked up and began to grow rapidly, their emissions likewise grew. And those increased emissions have completely swamped the (relatively small) emission reductions in the U.S. and Europe. Furthermore, on a per capita basis, CO_2 emissions in most of Asia (as well as Africa and Latin America) are still far below levels in the U.S. and Europe. Given that reducing emissions is costly, a relatively poor country like India would naturally object to being asked to make the same percentage reductions as a wealthy country like the U.S.

Of course, Figure 1.2 only shows historical emissions, and you might argue that what really matters is *future* emissions. Yes and no. Yes, because if global CO_2 emissions were cut in half during the next decade (never mind whether doing so is even remotely feasible), that would significantly slow down the buildup of the atmospheric CO_2 concentration, which would in turn slow down increases in the global mean temperature. Clearly the

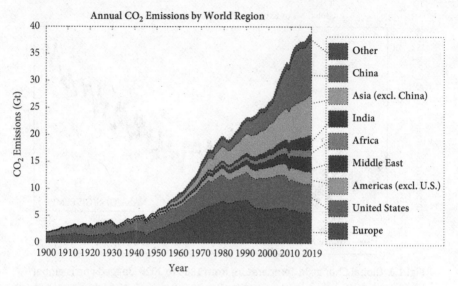

Fig. 1.2 CO_2 Emissions by Region, in billions of metric tons (gigatons). Since 1995, emissions from North America and Europe have declined somewhat, but emissions from Asia have been rising rapidly.
Source: Global Carbon Project, Supplemental Data, Global Carbon Budget, 2021.

CO_2 concentration and global mean temperature in, say, 2050, will depend considerably on emissions over the coming decades, and the policies that are adopted to reduce those emissions.

The problem is that the atmospheric CO_2 concentration has already grown enormously, as shown in Figure 1.3. The CO_2 concentration was around 300 parts per million (ppm) in 1950, and is now close to 420 ppm. Because CO_2 can stay in the atmosphere for well over 100 years, even if all further emissions were somehow cut immediately to zero, and remained at zero throughout the future, the atmospheric CO_2 concentration would remain above 400 ppm for decades to come. And because there is a two-decade or more time lag between increases in the atmospheric CO_2 concentration and its impact on temperature, even if we could *immediately* cut emissions to zero, temperatures will continue to rise because of *earlier* increases in the CO_2 concentration. (I will explain the nature of this time lag later; for now just take it as a given.)

I hope you will agree that CO_2 emissions cannot possibly drop to zero immediately. In fact, I will argue that under any realistic scenario, worldwide emissions will not drop to zero even over the next several decades. On the contrary, as Figure 1.2 shows, worldwide emissions have been growing rapidly over the past few decades, and there is no reason to think that this

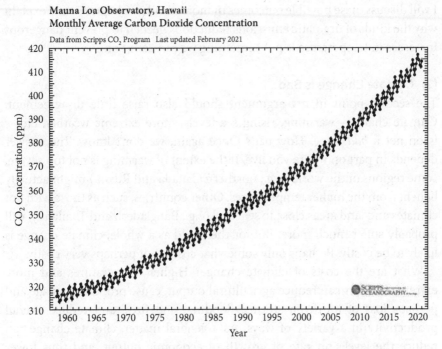

Fig. 1.3 Global Atmospheric CO_2 Concentration in parts per million (ppm). The sawtooth pattern is due to seasonal variation in the CO_2 level.
Source: Scripps Institution of Oceanography, www.scrippsco2.ucsd.edu.

growth will soon come to a sudden stop, and then change direction. As a result, we have to expect that the atmospheric CO_2 concentration will continue to increase over the coming decades. There is a good chance that by 2030, the CO_2 concentration will reach something close to 440 ppm. The sad fact is that past increases in the CO_2 concentration, along with likely future increases, will continue to push temperatures up, which in turn will lead to rising sea levels and more extreme weather.

You might say that I am just overly pessimistic. After all, the European Union has already pledged to reduce its net CO_2 emissions to zero by 2050, and China has recently stated its intention (not quite a pledge) to reduce its net emissions to zero by 2060.[1] But outcomes need not match pledges, even when those pledges are in the form of legal requirements. The U.K. government, for example, has adopted legislation (The "Climate Change Act") requiring the country to reduce its net CO_2 emissions to zero by 2050. But what happens if 2050 arrives and net emissions are still well above zero? Who goes to jail?

---

[1] Net emissions of CO_2 are emissions minus *removal* of CO_2 from the atmosphere, by planting trees or other means. I discuss carbon removal in the next chapter.

I will discuss these possible outcomes in more detail in Chapter 5, and explain why the kinds of dramatic emissions reductions needed to prevent dangerous increases in temperature is not something we can count on.

(2) Climate Change is Bad
The second point in my argument should also raise little disagreement: Climate change—warming, rising sea levels, more extreme weather, etc.— is on net a *bad thing*. How bad? Once again, we don't know. The answer depends in part on where you live. If the extent of warming is not too severe, some regions of the world (e.g., northern Canada and Russia) might actually benefit from the higher temperatures. Other countries, such as those with hot climates and land areas close to sea level (e.g., Bangladesh and Thailand), will probably suffer much more. But for the world as a whole, climate change is likely to be costly. Perhaps only somewhat costly, but perhaps very costly.

What are the costs of climate change? Higher temperatures and more extreme weather can reduce agricultural output, cause property damage and perhaps loss of life from storms, flooding, and fires, and reduce overall productivity in a variety of ways. As a general matter, climate change can reduce the level and rate of growth of economic output, and thus lower our standard of living. Because many harmful microbes and parasites thrive in warmer weather, and because very high temperatures can themselves be detrimental to health, it may also result in greater morbidity and mortality. And if it turns out to be severe, climate change could lead to social unrest and possibly even political upheaval.

Just how costly climate change is likely to be is very uncertain (for reasons I will explain later). You might think that the uncertainty implies that we should sit back (with a "Don't worry, be happy" attitude), at least until we learn more. Not so. On the contrary, the uncertainty itself gives us even more to be worried about, and reason to act sooner. That, too, will be explained later.

(3) We Need to Take Action
My third point is also (I hope) not very controversial: There is little disagreement that the world should take action to reduce the likelihood of severe climate change. But what kind of action? Almost all policy analyses of climate change and almost all policy recommendations focus on one basic kind of action: *sharply reducing GHG emissions*. Furthermore, it is generally agreed that nearly *all countries* must reduce their GHG emissions—not just relatively wealthy countries like the U.S., which currently accounts for only about

15 percent of worldwide emissions, but also developing countries like India, and the country that is now the largest single emitter of GHGs, China.

There are a variety of different ways to reduce emissions—a carbon tax or cap-and-trade system is the most direct (and generally considered the most economically efficient) way. But for now a carbon tax seems politically unpopular in the U.S. and many other countries, and it is not the only way to reduce emissions. So some policy analysts and politicians have instead (or in addition) proposed targeted regulations (e.g., automobile mileage standards) and subsidies for "green technologies" (such as solar and wind power, electric cars, and R&D directed at these and related technologies).

But whatever the specifics, the basic objective of climate policy has almost always been the same: *Reduce GHG emissions as much and as quickly as possible.* Yes, a good thing to do, but as I will explain, not enough.

(4) Reducing Emissions Won't Do the Job

So far, everything I've just said is more or less common knowledge, and unlikely to raise the hackles of many readers. But here is where things become a bit more controversial. I will argue that given any realistic scenario for worldwide GHG emissions over the coming decades (and even some unrealistic scenarios), *climate change is likely to be inevitable.* Yes, we should try hard to reduce emissions, and it is likely that many countries will indeed adopt policies that will lead to significant reductions in emissions. But it won't be enough.

Under any realistic scenario, even a highly optimistic scenario in which most countries agree to substantial emission reductions, atmospheric GHG concentrations will continue to grow for at least the next couple of decades. In fact, atmospheric GHG concentrations are already high enough to cause substantial increases in the global mean temperature. As a result of the high and growing atmospheric GHG concentrations, temperatures will continue to increase, although by how much and how soon we can't say. But some amount of warming—perhaps a great deal—is likely, as are the other aspects of climate change (rising sea levels, more extreme weather, etc.) that warming can bring about.

Some readers will surely take issue with this pessimistic assessment of our ability to reduce GHG emissions sufficiently to avoid significant warming. Why can't the U.S.—along with most other countries—adopt some kind of "Green New Deal" that involves extremely sharp reductions in emissions? Wouldn't an extension of the Paris Agreement (now with the participation of the U.S.) get us most of the way there? Doesn't my assessment of what might or might not be possible just boil down to simple defeatism? I think

I can convince most readers otherwise, although you'll have to read on and be patient.

It is important to stress at the outset that I am *not* suggesting we give up on reducing GHG emissions, or that reducing emissions is less important than we originally thought. On the contrary, we should push hard to reduce emissions, both locally (in each country) and worldwide. Ideally we would do this with a carbon tax, perhaps one that could be agreed upon as part of an international abatement policy, and that would apply to most major emitting countries. And we should explore other policy instruments for emission reductions. But at the same time, it is essential that we be realistic. As I will explain, given the hard constraints on what is technologically, economically, and politically feasible, emission reductions simply won't be enough to prevent climate change—in fact, not nearly enough.

The sad but fundamental problem is that over the coming decades worldwide GHG emissions are likely to grow, and atmospheric GHG concentrations will surely grow. This will be the case under any conceivable (but realistic) climate policy. Current targets for emission reductions vary, but even the more aggressive targets are insufficient to prevent increases in atmospheric GHG concentrations. (As part of the Paris Agreement, for example, China pledged to reduce the *growth rate* of its emissions between now and 2030, but it did not pledge to reduce its *level* of emissions.) Furthermore, it is very unlikely that the world will even come close to meeting current targets for emission reductions, never mind the more ambitious targets that have been proposed. Thus we must come to grips with the likelihood of a global mean temperature increase over the next 50 to 70 years that could turn out to be 3°C or even higher—well above the 1.5°C to 2°C that many climate scientists and policy analysts have argued is a critical limit. This could lead to rising sea levels, greater variability of weather, more intense storms, and other forms of climate change.

(5) The "Climate Outcome" Is Highly Uncertain

The literature on climate change, including books, articles, and popular press reports, have created the illusion that we know a great deal about climate change and its impact. Part of the problem is that large computer models have been developed and used to make projections, and those models convey an aura of scientific legitimacy and thus precision. But the problem is also that we humans prefer certainty to uncertainty, and feel more comfortable when we are given precise projections, for example that by 2050 the temperature will rise by X°C, sea levels will rise by Y meters, and as a result GDP will fall by Z percent. But projections of that sort can be highly misleading. The fact is that even if we could accurately predict future GHG emissions, we don't know—and at this point can't know—by how much the temperature or sea levels will rise. And even if we could accurately predict how much the temperature and sea levels will rise, we don't know what the impact would be on GDP or other measures of economic and social welfare.

The basic fact is that the "climate outcome," by which I mean the extent of climate change and its impact on the economy and society more generally, is highly uncertain, much more uncertain than most people think. I'll explain the extent of the uncertainty and the reasons for it in detail in Chapter 3. But for now, take my word for it: We don't know how extensive and severe climate change will be, and even if we could predict the extent and timing of future climate change, we know even less about what its impact will be.

Suppose that by 2050 or 2060, the global mean temperature increases by 3°C. That would be a larger and more rapid temperature increase than the central forecasts made by the Intergovernmental Panel on Climate Change (IPCC) and others, but it is certainly within the range of what they consider possible. By how much would sea levels increase, and how much flooding would result in various parts of the world? To what extent would the severity and frequency of droughts and storms increase? What would be the impact on agriculture, and on economic activity generally? The answers to all of these questions are that we don't know. It might turn out that however severe climate change turns out to be, its impact will be mild or moderate. But there is also a possibility that we will not be so lucky, and that we will endure climate change impacts that are severe, even catastrophic—especially if society is unprepared.

Hopefully we will learn more as research in climate science continues, and the uncertainty will be reduced. But that will take time, and right now the uncertainty is considerable. What are the implications of all of this uncertainty? Doesn't this mean that we should wait and see what happens? After all, if we don't know how much the climate will change, and we don't know what the impact of climate change will be, why take costly actions now?

That is indeed the argument made by many of the people who oppose carbon taxes or other measures to reduce emissions.[2] But that argument is wrong, and gets things backwards. The fact is that *the uncertainty itself is a reason to act now.* You don't know whether your home will be damaged by a fire, flood, or a falling tree during the next several years, never mind

---

[2] In Pindyck (2013a), I argued that integrated assessment models (IAMs) "have crucial flaws that make them close to useless as tools for policy analysis" (page 860), and that those models can create a perception of knowledge and precision that is illusory. That paper, and a follow-on (Pindyck (2017b)), have led some people to label me as a "climate denier," even though I made clear in those papers that the uncertainty does not mean we shouldn't move aggressively to deal with climate change.

how much damage might result from such an event. But that doesn't mean you shouldn't buy insurance for your home. On the contrary, a prudent homeowner will buy enough insurance to cover the potential cost of an adverse event. Likewise, we don't know what the future costs of climate change might be, but that doesn't mean we should ignore the problem and take no action. Instead, we should take action now as insurance against the possibility of very high costs in the future.

(6) Invest in Adaptation

If I am right about the risk of a severe climate change outcome, and the importance of taking action now, what should we do beyond trying to reduce CO_2 emissions as much as possible? Once again, here is where I depart from what is generally considered common knowledge. I will argue that it is essential to insure against the possibility of a catastrophic climate outcome, and given that emission reductions won't be sufficient, the best thing to do is to invest now in *adaptation*.

Adaptation means taking steps to counter the warming effects of a high and rising CO_2 concentration, or any of the other aspects of climate change that warming can bring about. As I will explain in detail below, adaptation can have many forms—developing new hybrid crops that can resist high temperatures, adopting policies to discourage building in flood-prone or wildfire-prone areas, building sea walls and dikes to prevent flooding, and forms of geoengineering that can reduce the greenhouse effect of a rising CO_2 concentration are examples. Developing new ways to abate emissions remains important and should be pursued vigorously, but climate change research, and climate change policy, should put much more emphasis on adaptation than has been the case so far.

Wouldn't adaptation along any of these lines be prohibitively expensive? No, adaptation need not be very expensive, for reasons that I'll lay out later in this book. In fact, in many cases adaptation is much cheaper than reducing emissions, and certainly cheaper than *sharply* reducing emissions. Yes, the costs of alternative energy sources such as wind and solar are falling dramatically, our cars and airplanes are becoming much more fuel efficient, we are able to insulate our homes and buildings more effectively, and lighting, refrigeration and air conditioning have all become more energy efficient. Those are some of the reasons that we will be able to reduce GHG emissions by 20, 30, or even 40 percent over the coming decade or two, and at a reasonable cost. But reducing emissions by 80 percent? Yes, it can be done, but it would be extremely costly. And adaptation? As you will see (if you are patient and keep reading), the costs can be far lower.

That summarizes my basic argument. We must come to grips with the likelihood of a global mean temperature increase over the next 50 years well above the 2°C "limit." The temperature increase might turn out to be 3°C, or even higher. This degree of warming could lead to rising sea levels, greater variability of weather, more intense storms, and other forms of climate change. The extent and impact of such climate change is highly uncertain; we might be lucky and experience only a mild or moderate climate outcome, but the outcome could instead be catastrophic, especially if society is unprepared. The best way to prepare is to invest now in adaptation. Developing and implementing new ways to abate emissions remains important and should be pursued aggressively. But at the same time we need to come to grips with the fact that emissions abatement won't be enough. Thus climate change research, and climate change policy, should put much more emphasis on adaptation.

Now, a little more detail on what is meant by adaptation.

1.2 What Is Adaptation?

Adaptation is certainly not a new idea. In fact, under the Paris Agreement, steps to deal with the impact of climate change receive considerable attention. Just as parties to the Agreement will submit plans for reducing GHG emissions, the Agreement requires all parties to plan and implement adaptation efforts, and encourages parties to report on those efforts. However, the Paris Agreement is vague about what is included under "adaptation."

There are two different kinds of adaptation, both of which can reduce the threat of climate change. The first kind of adaptation involves doing things that will *reduce the harmful impacts of climate change*, but not prevent climate change from happening. How can we reduce the impacts of climate change, especially after it has occurred? Installing air conditioning, improving the ability of our public health systems to respond to heat waves, developing hybrid crops that are more resistant to heat and drought, changing where and how we build homes, and building sea walls to prevent flooding from higher sea levels are examples.

The second kind of adaptation involves doing things to *reduce the warming effects of a rising CO_2 concentration*, as opposed to stopping or slowing the rise. Reducing the warming effects of atmospheric CO_2 would reduce the extent of climate change itself. But how can we reduce the warming effects of a rising CO_2 concentration? It can be done through forms of *geoengineering*, and most notably solar geoengineering. I'll explain how geoengineering works in detail later, but one basic idea is to inject sulfur (or some other material) into the atmosphere, using airplanes or balloons. "Seeding" the atmosphere in this way would reflect some sunlight away from the earth and thereby reduce the greenhouse effect, which in turn would reduce the amount of warming. It would not eliminate CO_2 from the atmosphere, but would render that CO_2 much less harmful.

I will discuss these and other examples of adaptation below, but first we have to address the question of *who does the adaptation*, and distinguish between *private* versus *public* adaptation. Private adaptation involves actions taken by households and private firms. Public adaptation involves actions taken by local, state, and federal governments. And sometimes adaptation can involve a mix of private and public actions.

Private Adaptation

What actions can households and private firms take to adapt to climate change? My guess is that many readers have already taken some of these actions. For example, you might have installed air conditioning in your home to reduce the impact of higher temperatures. Or you might have forgone the possibility of building or buying a beach-front home, because such a home would be more vulnerable to rising sea levels and stronger hurricanes.

To some extent, private firms have also started to respond to climate change. Real estate developers are now less likely to undertake the construction of beach-front houses and condominiums. The construction of retirement homes and communities in northern states is beginning to look more attractive than in some parts of Florida. As the demand for air conditioners rises, firms like Carrier Corporation are investing in greater production capacity, and in the development of cheaper and more efficient units. And agricultural biotechnology companies like Pioneer Hi-Bred International and the Swiss firm Syngenta have used conventional breeding techniques to develop drought-tolerant corn varieties for use as animal feed. Likewise, Monsanto has developed genetically modified strains of corn and rice that are more resistant to heat and drought, and they continue to do research on the heat-resistant grains.

So far climate change has been limited—temperatures have risen, but not by much—so adaptation by households has likewise been limited. But we can expect that as temperatures climb and weather conditions become more variable and extreme, the extent of private adaptation will increase.

Public Adaptation

Although households and private firms can contribute to climate change adaptation, some forms of adaptation will require the government to play a major role. The reason is that some of the most effective tools for adaptation involve large-scale projects that go far beyond the capacities of private firms (never mind households). Here I will briefly discuss two examples, both of which will get much more attention later in this book.

The first example is sea walls, and more generally flood walls, dikes, levees, and other barriers to prevent flooding from rising sea levels. If sea levels rise by several feet (which is a substantial amount), it will not immediately cause flooding, at least in most areas. The problem is that a higher sea level makes coastal areas more susceptible to flooding from storm surges, i.e., from high waves produced by strong storms and hurricanes. An example was the flooding of southern Manhattan during Hurricane Sandy on October 29, 2012. The hurricane had winds of 80 mph, which produced waves that inundated large areas of the city, including the subway system, the tunnels going in and out of the city, and a large number of buildings.

Remember that Hurricane Sandy happened *before* any significant warming-induced rise in sea levels. Should sea levels rise over the coming decades, we can expect storm surges like this to be stronger and more frequent, and to cause even worse flooding. So what can be done? One proposal is to build a sea wall around Manhattan, along the lines of the one illustrated in Figure 1.4. (In fact, in 2016, $176 million in federal funding was allocated for the project, largely to pay for further study. That study led

Fig. 1.4 Proposed sea wall around southern Manhattan that would prevent flooding from a storm surge like the one that occurred during Hurricane Sandy in 2012. $176 million in federal funding was allocated for the project in 2016, and later the plan was expanded to nearly $1 billion.

to a more expensive plan expected to cost several billion dollars.) A sea wall like this need not extend above the waterline, and depending on the design, could be invisible from the shore. Its purpose is not to prevent flooding from a normal sea level, but rather to prevent flooding from surges.

Sea walls are not a new idea. Nor are dikes, levees, and other barriers to flooding. Much of the Netherlands, for example, is below sea level, but is protected from flooding by a large network of dikes. The earliest dikes in the Netherlands were built some 800 years ago, and construction and structural improvement has continued since. The question, discussed in detail later in this book, is whether sea walls and dikes can provide an economical form of public adaptation to climate change.

Sea walls help to reduce the harmful impacts of climate change. The second example is quite different: *solar geoengineering*, which would reduce the warming effects of atmospheric CO_2. We'll discuss this in detail later in the book, but the basic idea is quite simple: "Seed" the atmosphere, at an altitude of roughly 60,000 to 80,000 feet, with sulfur or sulfur dioxide. These "seeds" would remain in the atmosphere for up to a year, after which time they would precipitate as sulfuric acid and fall back to earth. (Thus the "seeding" would have to be repeated regularly.) While in the atmosphere the particles would reduce the greenhouse effect by reflecting sunlight back into space. The CO_2 will remain in the atmosphere, but the sulfur dioxide will nullify much of its harmful warming effects.

Solar geoengineering might seem expensive, but it's not. Yes, the sulfur dioxide will eventually come down from the atmosphere in the form of sulfuric acid, so that the "seeding" will have to be repeated at least every year or so. But the cost of the seeding itself is actually quite low. And that low cost creates another advantage: It partly eliminates the free-rider problem that makes emissions abatement so difficult. As I explained earlier, and was illustrated in Figure 1.2, the U.S. and Europe are already reducing CO_2 emissions, but not so most of the Asian (and other) countries. Rather than reducing its own emissions (at considerable cost), a country like India can "free ride" on the emission reductions of other countries. But because it is so cheap, solar geoengineering doesn't require the participation of all or even most countries—it could be done, and done effectively, by just a few countries.

As we will discuss in more detail later, solar geoengineering is not a simple cure-all, and in fact it is highly controversial. For example, it might create its own environmental problems, in part because CO_2 will continue to accumulate in the atmosphere, and some of it will be absorbed by the oceans, making them more acidic. In addition, the sulfuric acid that eventually rains down can make lakes and rivers more acidic. But putting these concerns aside for now, geoengineering is an important option for adaptation.

Public and Private Adaptation

Adaptation can also involve a mix of private and public actions. Take, for example, our concern about rising sea levels, which could wash away beach-front homes. The decision to build a home on or near the ocean is a private one, but it is influenced by government policy: Will the government provide at least partial insurance should the home be destroyed in a major hurricane? Currently the U.S. government does provide such insurance, which means that in effect we—society—are subsidizing the cost of beach-front homes. A possible public action would be to eliminate or reduce such insurance, which would lead to the construction and sale of fewer beach-front homes.

A second public-private example is the development of heat-tolerant strains of wheat, corn, and other grains. As I mentioned above, private agricultural firms are already working on this, but so, too, are state and federal governments. For example, the U.S. Department of Agriculture conducts research on crops (and food supplies generally), and also supports research done by other organizations (such as universities).

Governments can and do provide incentives to encourage homes and businesses to reduce GHG emissions, e.g., by subsidizing electric vehicles, energy sources such as solar and wind, and R&D to develop new "green" technologies. But governments can likewise provide incentives for homes and businesses to adapt to climate change, e.g., by subsidizing the purchase and installation of drains and sump pumps to reduce flood risk.

Resilience

As I will stress repeatedly, we face a lot of uncertainty when it comes to climate change. We don't know exactly what will happen to temperature, sea levels, and hurricane intensity, nor do we know what the possible impacts of those changes might be. Furthermore, many of the impacts will be *local* in nature; Miami is likely to experience much greater damage from rising sea levels than, say, Denver, or even other coastal cities such as Los Angeles or Boston. As a result, when possible adaptation should improve our *resilience* to climate change. Sea walls and levees are an example of this; they would make a city (Miami, let's say) much more resilient to possible (and largely unpredictable) storm surges.

Resilience is especially important in developing countries. Will climate change result in more droughts or much more rain in a country like Ethiopia? At this point we don't really know. Paving rural roads might therefore be a good form of adaptation—if there are more floods, it will still be possible to bring crops to market, and if not, the paved roads will bring other economic benefits.

Amelioration vs. Adaptation

Some discussions of climate change distinguish between adaptation and *amelioration*. The idea is that adaptation refers to things that reduce the damages from whatever climate change actually occurs, whereas amelioration refers to reductions in warming and other forms of climate change, despite an increase in the atmospheric CO_2 concentration.[3] Thus sea walls would fall into the category of adaptation because they reduce the flooding caused by rising sea levels, but solar geoengineering would fall into the category of amelioration, because it would reduce the warming effects of increases in the atmospheric CO_2 concentration.

I don't find this distinction to be very useful. What matters is the distinction between policies designed to reduce the amount of CO_2 emissions, which is what most discussions of climate policy focus on, and policies designed to reduce the harmful impacts of CO_2 emissions. Thus I will simply refer to sea walls and solar geoengineering as different kinds of adaptation.

The critical point is that any feasible reductions in CO_2 emissions will not be sufficient to eliminate the risk of catastrophic climate change. We also need to do other things, all of which I consider to be forms of adaptation.

1.2.1 Concerns about Adaptation

Assume for the moment that you agree with my assertion that under any realistic scenarios for climate policy, atmospheric GHG concentrations will continue to grow, making climate change inevitable. You might still be very unhappy with my proposed focus on adaptation as a solution, even a partial solution, to the problem. Why? For several possible reasons.

First, you might argue that most forms of adaptation are too speculative to rely on. How do we know that a sea wall around lower Manhattan can actually protect the city from the storm surges that could be larger and more frequent because of rising sea levels and stronger hurricanes? And how do we know that new crops can be developed that will be able to withstand more extreme temperatures? And do we know that solar geoengineering strategies designed to reverse the greenhouse effect and thereby reduce warming will

---

[3] In a recent paper, Aldy and Zeckhauser (2020) make this distinction.

actually work?[4] We don't know with certainty how well these and other forms of adaptation will work. However, as I will explain later, our understanding and experience with the technologies behind geoengineering, sea walls, new crops, and other forms of adaptation give us considerable confidence that the technologies *will* work, and can be implemented at reasonable costs.

Second, you might argue that some forms of adaptation will also cause environmental damage. Perhaps the most obvious example is air conditioning, the simplest form of private adaptation to higher temperatures. The demand for air conditioners in India is expected to explode, but as long as much of the electricity needed to run those air conditioners is generated from fossil fuels, that will push up CO_2 emissions.

There are plenty of other examples of possible environmental damage from adaptation. Will the sulfur dioxide used in solar geoengineering, which eventually returns to the earth in the form of sulfuric acid, cause acidification of lakes and rivers? Might new crops, especially ones that have been developed via genetic modification, turn out to be harmful "Frankenfoods?" Yes, these are indeed concerns, and they need to be addressed. But as I'll explain, they are small compared to the concerns we have (or should have) about climate change itself. And in most cases, the risks behind these concerns can be managed.

Third, won't a focus on adaptation deflect us from doing what we should be doing to reduce GHG emissions? If we accept the view that we can adapt to climate change, why bother with costly measures to reduce emissions? This is a good point, and probably the most common objection to adaptation. Indeed, it is the reason that many environmentalists view the very word "adaptation" as anathema. There is certainly some truth to this concern— and remember that I am not arguing that we should give up on reducing emissions. If we know that there is a cheaper and easier alternative, why go to the effort and considerable expense of reducing emissions? But there is another side to that coin: If we can achieve the same objective with adaptation far more cheaply than with emissions abatement, why forego that lower-cost option? Shouldn't it be something that we at least consider *in addition* to reducing emissions?

That brings me to the last objection to adaptation: the claim that humans have no right to interfere in any way with our natural environment. No right to interfere with the environment by producing GHG emissions, but also no

[4] In a recent article Nordhaus (2019) argues that geoengineering is "untested, will not offset climate change equally in all regions, will not deal with ocean carbonization, and will have major complications for international cooperation."

right to interfere with the environment via solar geoengineering, sea walls, or the development of "artificial" crops. Of course our very existence interferes with the natural environment, but adaptation would go beyond that. This is effectively a philosophical or religious argument, which I can't really respond to, because I'm an economist, not a philosopher or theologian. So I will stick to the economic aspects of climate change and climate policy.

1.2.2 Carbon Removal and Sequestration

I have argued that under any realistic scenario for emission reductions, the atmospheric CO_2 concentration will continue to rise, pushing up temperatures. But perhaps I've missed something. Can't we *remove* CO_2 from the atmosphere, and thereby counter the increase in concentration?

Indeed, another proposed approach to deal with the build-up of CO_2 in the atmosphere is to remove some of it (*carbon removal*), and then store it in a way that will prevent its future release into the atmosphere (which is called *carbon sequestration*). Carbon removal and sequestration certainly sounds attractive, and in principle would have no negative environmental impact. Removing carbon from the atmosphere is not really a form of adaptation, but instead should be thought of as a way to reduce "net" emissions (i.e., total emissions minus the CO_2 removed), or equivalently an "undoing" of the growing CO_2 concentration. I'll discuss carbon removal and sequestration in more detail later in this book, but here is the key question: Might it provide a realistic solution to the climate change problem?

One obvious way to try to remove CO_2 is to plant trees, which is seen as a tool for climate policy in some countries. Trees (and other green plants) grow by absorbing CO_2 and combining it with water and the energy from sunlight. In the process, oxygen is emitted. So, more trees means more absorption of atmospheric CO_2, and lower net emissions.

In fact, one of the factors that has contributed to the gradual increase in the atmospheric CO_2 concentration is *deforestation*. During the past decade, roughly 1 billion trees have been cut down each year in the Amazon Rain Forest alone, and many more have been cut down and burned to make room for palm oil plantations in Indonesia and Malaysia. (Burning the trees releases even more CO_2.) At the very least we should sharply reduce or even halt deforestation, but right now that does not seem likely. What about planting new trees? As I will discuss in Chapter 6, planting trees would indeed reduce the atmospheric CO_2 concentration somewhat, but it would take a very large number of new trees to make much of a difference. Furthermore, trees require land and water, both of which are expensive, making it unclear where and how those new trees would be planted.

What about other forms of carbon removal, such as absorbing, sequestering, and storing the CO_2 as it is produced in fossil fuel burning power plants? There have been proposals to do just that, which I will discuss in detail in Chapter 6. But for now the technologies involved are very expensive, currently much too expensive to make economic sense.

The bottom line is that unfortunately carbon removal and sequestration on any large scale has a big problem: We don't know how to do it, at least not at anything approaching a reasonable cost. We simply lack the required technology. Perhaps a technological breakthrough will happen in the coming decade or two, but right now carbon removal and sequestration on a significant scale is far too expensive to be considered a serious solution to climate change.[5] Nonetheless, carbon removal and sequestration is seen by many as a potentially important tool, and emission targets are often expressed in terms of "net" emissions. It may indeed help, but barring a major technological advance, carbon removal and sequestration won't do much to reduce the accumulation of CO_2 in the atmosphere. At least it's not something we can count on now.

1.3 What Comes Next

In this Introduction I summarized the basic argument behind this book: While reducing GHG emissions as much and as quickly as possible should be an important part of climate policy, under any realistic scenario it is very unlikely that we will be able to prevent the global mean temperature from rising more than 2°C by the end of the century. In the next chapter I will illustrate this problem by presenting a simple calculation based on a very optimistic scenario for emission reductions.

Of course what matters is the *impact* of higher temperatures, not the higher temperatures themselves. There is considerable uncertainty over what the impact of higher temperatures will be, but the impact could be severe. That means we will need another way to deal with climate change, namely adaptation. I presented a few examples of adaptation, but they are just examples, and we need to look at the range of adaptation options in more detail.

---

[5] For more detail on carbon removal, see National Research Council (2015). Also, regarding deforestation of the Amazon rain forest, see Franklin and Pindyck (2018) and the references therein.

So what comes next in this book? In the next chapter, which you can think of as "Climate Change Basics," I introduce some important terms and concepts that are used frequently in discussions of climate change and climate policy. I'll explain how we measure things (temperature, CO_2 emissions, the atmospheric CO_2 concentration, etc.), and present a few basic facts about how climate change occurs. That will allow me to present some simple calculations of the temperature change we might expect over the rest of this century, based on a very optimistic scenario for emission reductions.

Even if we knew exactly how much CO_2 and other GHGs will be emitted into the atmosphere over the coming decades, there is considerable uncertainty over what the impact would be on temperatures, and (indirectly) on other measures of climate change. And even if we knew exactly how much the global mean temperature would rise over the coming decades, there is considerable uncertainty over the impact that warming and rising sea levels would have. Given all of the research that has been done over the past few decades, why is there still so much uncertainty over climate change and its impact? To address that question, I need to spend some time explaining in more detail what we know and what we don't know about climate change, where our knowledge is lacking, and how some of the uncertainty might or might not get resolved in the future. That will be the subject of Chapter 3.

In Chapter 4, I turn to the implications of all this uncertainty for climate policy. Shouldn't the uncertainty lead us to slow down on climate policy, rather than take costly actions now? As I'll explain, the uncertainty affects policy in two ways. First, it creates *insurance value*: Acting early to reduce GHG emissions reduces the likelihood of a catastrophic climate outcome. Second, it raises the importance of *irreversibilities*: Waiting to do anything causes a (nearly) irreversible increase in the atmospheric CO_2 concentration, whereas acting early implies irreversible expenditures on emissions abatement. As we'll see, these irreversibilities work in opposite directions.

Chapter 2 presents some very simple calculations of the temperature changes we might expect over the coming century, even if we succeed in reducing CO_2 emissions sharply. I return to these calculations in Chapter 5, beginning with what we might expect—even if we are die-hard optimists— in the way of emission reductions over the coming decades. For example, what kinds of paths for global CO_2 emissions might be feasible? And what would those paths imply for changes in the global mean temperature up to the end of the century? I will address these questions using a simple model of both CO_2 and methane emissions, atmospheric accumulation, and impact on temperature. (A simple model, but a bit more complicated than what I used for the calculations in Chapter 2.) I will show that there is no realistic emissions scenario that can eliminate the likelihood—or at least the strong possibility—of a temperature increase above 2°C by the year 2100.

In Chapters 6 and 7, I address the all-important question of what to do. In Chapter 6 the focus is on how we might go about reducing emissions. Should we rely on a carbon tax, and if so, how large should the tax be? Would a "cap and trade" system be preferable, and how might it work? To what extent should we rely on a mix of direct regulations and subsidies for "green" technologies? Given that climate change is a global problem, how can we reach an international agreement that avoids the free-rider problem, by which all countries reduce emissions? And what should be the role of nuclear power as a way of "decarbonizing" the production of electricity? Finally, I discuss two very different ways to remove CO_2 from the atmosphere and thereby reduce *net* emissions: planting trees, and carbon removal and sequestration.

Then the question is what do to beyond reducing emissions, and that will take me to the role of *adaptation*, the subject of Chapter 7. I will review different forms of private and public adaptation, and discuss several examples in detail. I explain how agriculture—highly vulnerable to extremes in temperature and rainfall—has already been adapted to climate change, and is likely to see further adaptation. I also discuss the potential for adaptation to rising sea levels and more frequent and powerful hurricanes. And then I turn to what may be the most important—and most controversial—form of adaptation, which is geoengineering.

1.4 Further Readings

This book is not intended to be an introduction to the science and/or economics of climate change. Instead, the goal is to explore aspects of climate policy, and to explain how and why current thinking about climate policy is misguided. Although I explain in the book what we know and don't know about climate change, my discussion is fairly brief, and some readers might like a more detailed introduction to the subject. For those readers, I would suggest the following books and articles:

- *Climate Shock* by Wagner and Weitzman (2015) provides a nice introduction to the science and economics of climate change, and explains the nature of the uncertainty regarding what we might expect. Given the possibility of an extreme outcome, it emphasizes the importance of "radical" forms of adaptation, an example of which is geoengineering.

- In *The Climate Casino*, Nordhaus (2013) uses his DICE (Dynamic Integrated Climate and Economy) model to help explain—at a textbook level—how unrestricted GHG emissions can cause climate change to occur, and can lead to serious problems in the future. He also utilizes the model to illustrate some of the uncertainties we face when thinking about the climate system and when trying to predict the changes to expect under different policies. The book thereby provides students (and others) with a good introduction to climate change policy. Also, Nordhaus (2019) provides a nice article-length overview of the economics of climate change, and why climate change policy is so important.

- Three other books that provide nice introductions to climate change, with more of a focus on the science, are *Climate Change: What Everyone Needs to Know* by Romm (2018), *Global Warming: The Complete Briefing* by Houghton (2015), and *What We Know about Climate Change* by Emanuel (2018).

- Others have also argued that reducing CO_2 emissions will not be sufficient to eliminate the chance of a catastrophic climate outcome, and that adaptation is also needed. See, for example, Aldy and Zeckhauser (2020) and Keith and Deutch (2020).

- And lastly, for a detailed discussion of the climate change problem, with conclusions similar to those in this book, read Gollier (2019), Christian Gollier's beautifully written recent book *The Climate after the End of the Month*. Yes, it's in French, but that shouldn't be a problem. (Why *after* the end of the month? Because your paycheck arrives at the end of the month, and that—unfortunately—is more important than climate change for most people.)

2

The Fundamental Problem

The basic mechanism behind climate change is not that difficult to understand. When energy in the form of sunlight reaches the Earth's atmosphere, part is reflected back into space and the rest is absorbed by the planet. In addition, some energy is always radiating away from the (relatively warm) Earth into (relatively cold) space. The difference between the energy flowing in and the energy flowing out is called *radiative forcing*, and if that difference is positive (more energy is flowing in than flowing out), the Earth will get warmer. Atmospheric CO_2 causes radiative forcing by increasing the fraction of sunlight absorbed relative to the fraction reflected into space, thus warming the planet.

As we burn carbon, more and more CO_2 accumulates in the atmosphere. (Other GHGs are emitted as well, most notably methane. We'll get to those later, but for now I'll focus only on CO_2.) Higher concentrations of CO_2 increase the amount of radiative forcing. In other words, the CO_2 creates what we call a "greenhouse" effect, trapping more heat (from the sun and from the Earth's surface) and thereby causing an increase in temperatures. Higher temperatures cause other changes in the environment and the climate. For example, warming can cause sea water to expand and cause glaciers and ice sheets in Greenland and Antarctic regions to break apart, raising sea levels and possibly inundating coastal areas. And higher ocean temperatures contribute more energy to tropical storms and hurricanes, making them more intense and more destructive.

The problem is predicting *how much* temperatures will increase and sea levels will rise, and what the impact of those changes might be. There is a lot of uncertainty here, which we will get to in the next chapter. But for now let's return to the claim I made in the Introduction—that under any realistic emissions scenario, the CO_2 concentration in the atmosphere is likely to rise substantially, so that temperatures will likewise rise, and we will need to rely on adaptation.

At this point you may be very skeptical about this claim. You might think that I am simply a defeatist, and turning to adaptation means turning away from the job we need to do to reduce emissions. Maybe you think that a

Climate Future: Averting and Adapting to Climate Change. Robert S. Pindyck, Oxford University Press.
© Oxford University Press 2022. DOI: 10.1093/oso/9780197647349.003.0002

"Green New Deal" and something along the lines of the Paris Agreement will do the job and save us from climate change. (In fact, the underlying assumption of the 2015 Paris Agreement is that countries will agree to emission reductions sufficient to prevent the global mean temperature from rising 2°C by the end of the century.) So let's return to this fundamental question: Why can't reducing CO_2 emissions—aggressively—prevent a temperature increase above 2°C? To address this question, we need to discuss some of the details about how CO_2 emissions result in higher temperatures. And that means we need to get into the numbers.

2.1 A Few Facts and Numbers

Later I will provide a detailed analysis of the temperature implications of alternative CO_2 emission trajectories, such as when emissions might stop growing, when and how fast emissions might decline, etc. But in this chapter I'll try to illustrate the problem with a some simple "back of the envelope" calculations. However, before I can do that, I have to lay out some basic information about how we measure things like emissions and CO_2 concentrations, and how an increase in the atmospheric CO_2 concentration leads to higher temperatures.

First, how do we measure things? Here is what you need to know:

(1) **Temperature:** In the climate change business we usually measure temperatures (and temperature change) in terms of degrees Celsius, as opposed to degrees Fahrenheit. So a temperature increase of 1.0°C means 1.0 degree Celsius, which is equivalent to 1.8 degrees Fahrenheit (1.8°F).

(2) **CO_2 Emissions:** We measure quantities of CO_2 in terms of metric tons. (A metric ton is equal to 1000 kilograms, or for those of you averse to the metric system, roughly 2205 pounds.) Annual CO_2 emissions are measured in billions of metric tons, called *gigatons* (Gt) for short. As shown in Figure 1.2 in the previous chapter, worldwide CO_2 emissions in 1950 were approximately 6 Gt, but by 2019 annual emissions had increased to 37 Gt.

(3) **Net Emissions:** As I will explain later, there are ways to remove CO_2 from the atmosphere ("carbon removal") and store it to prevent its re-entry ("carbon sequestration"). Planting trees is one example of how this can be done. To the extent that carbon removal and sequestration is feasible, we would want to measure CO_2 emissions net of the amount removed from the atmosphere. Some countries (e.g., Britain) have

adopted a "net-zero" emissions target for the year 2050, meaning that any CO_2 emissions in that year would have to be offset by an equal amount of CO_2 removal.

(4) **Carbon vs. Carbon Dioxide:** Sometimes you will see emissions measured in terms of tons of carbon, rather than tons of CO_2. A ton of CO_2 contains only 0.2727 tons of carbon (the rest of the weight comes from the two oxygen atoms bound to each carbon atom). So the 37 Gt of CO_2 emissions in 2019 is equivalent to $(37)(0.2727) = 10.1$ Gt of carbon. To avoid confusion, the numbers I present in this book are always in terms of CO_2, not carbon, unless indicated otherwise.

(5) **Atmospheric CO_2 Concentration:** The concentration of CO_2 in the atmosphere is measured in terms of *parts per million* (ppm). During pre-industrial times, around 1750, the atmospheric CO_2 concentration was around 280 ppm. The concentration rose slowly during the 19th and early 20th centuries, and in 1960, it was about 315 ppm. But since then the concentration has been rising steadily, and by 2020 it had increased to 415 ppm, as shown in Figure 1.3.

(6) **CO_2 Emissions vs. CO_2 Concentration:** Some CO_2 emissions are absorbed by the oceans ("ocean uptake," discussed below), and the rest remain in the atmosphere. Adding one Gt of CO_2 to the atmosphere will increase the CO_2 concentration by 0.128 parts per million (ppm). In 2019, global CO_2 emissions were about 37 Gt. Ignoring ocean uptake, that year's emissions increased the atmospheric CO_2 concentration by about $(37)(0.128) = 4.74$ ppm.

Next, a few facts, definitions, and more numbers describing how changes in the CO_2 concentration affect temperature:

(1) **Global Mean Temperature:** We often refer to warming in terms of increases in the *global mean temperature*, i.e., the planet's surface temperature averaged across locations around the world. But be careful, because temperature increases can differ considerably from one region of the world to another. For example, the increase in the global mean temperature since 1960 is already close to 1.0°C, but in the Arctic region the temperature has increased by about 1.5°C, and in the Southern Hemisphere by just over 0.5°C. Global mean temperature, however, is still a useful summary statistic that is used widely, and we will use it in what follows.

(2) **Climate Sensitivity:** A doubling (100 percent increase) of the atmospheric CO_2 concentration will cause an increase in temperature. No

surprise there, but *how much* of an increase? We're not sure, but the current state of climate science puts the number at something between 1.5 and 4.5°C. Yes, that's a wide range, and I'll explain why it's so wide in the next chapter. A commonly used number is 3.0°C, which is right in the middle of that range. That number—the increase in the global mean temperature resulting from a doubling of the atmospheric CO_2 concentration—is referred to as *climate sensitivity*. In the calculations below, I will use that 3.0°C number to translate increases in the CO_2 concentration to increases in temperature. For example, that number tells us that a 10 percent increase in the CO_2 concentration will cause a 0.30°C increase in the global mean temperature.[1]

(3) **Time Lag:** The temperature increase resulting from an increase in the atmospheric CO_2 concentration doesn't happen overnight. How long does it take? It depends in part on the size of the increase in the CO_2 concentration—the larger the increase, the longer is the time lag. For a small increase, the lag will be only about 10 years, but for a very large increases it can be as long as 40 or even 50 years. However, for any particular increase in the CO_2 concentration, there is some uncertainty over the time lag. So what number should we use when we calculate the temperature implications of alternative CO_2 emission trajectories? Climate scientists may disagree, but 30 years is a commonly used number. In what follows I will assume that the full effect on temperature of an increase in the CO_2 concentration takes 30 years.[2] In other words, if the CO_2 concentration suddenly doubled, it would have a small immediate effect on temperature, cause a 1.5°C temperature increase after 15 years, and a 3.0°C temperature increase after 30 years.

(4) **Dissipation of CO_2 from the Atmosphere:** Once it's in the atmosphere, CO_2 stays there a very long time. The rate at which CO_2 dissipates from the atmosphere is somewhere between 0.25 percent to 0.50 percent per year. So even the upper end of the range (a dissipation rate of 0.50 percent per year) means that about 78 percent of any CO_2 emitted today will still be in the atmosphere 50 years from now. In the scenario examined below, I will use a dissipation rate of 0.35 percent per year, which gives the best fit to the data from 1960 onwards.

---

[1] This is an approximation, but a reasonable one. The temperature change is actually proportional to the logarithm of the change in the CO_2 concentration, as opposed to the change itself. Also, in its most recent publication, the IPCC has narrowed the "most likely" range for climate sensitivity to 2.5°C to 4.0°C.
[2] The *full effect* can take more than a century, but most of the effect happens in a couple of decades or so. Thus the 30-year number is an approximation, but a reasonable one.

A 0.35 percent dissipation rate means that about 84 percent of any CO_2 emitted today will remain in the atmosphere 50 years from now, and 70 percent will still be there 100 years from now.[3]

(5) **Ocean Uptake:** When CO_2 dissipates from the atmosphere, where does it go? Most goes into the ocean ("ocean uptake"), and some into the land. The process of ocean uptake is complex, because depending on the relative air and ocean temperatures, the oceans are also releasing CO_2. As a result, the dissipation rate is not constant, so setting it at 0.35 percent per year is an approximation (but a reasonable one).

(6) **Ocean Acidification.** Ocean uptake creates an additional problem: The absorption of CO_2 makes the oceans somewhat more acidic, and there is concern that this could adversely affect coral and other ocean ecosystems. (I discuss ocean acidification and its possible impact later, in Chapter 7.)

(7) **What Goes Up Need Not Come Down:** Suppose the world could somehow dramatically reduce CO_2 emissions. In fact, suppose CO_2 emissions are cut to zero within the next few years. What would happen? Well, once emissions are eliminated, the atmospheric CO_2 concentration will stop growing, and will begin to *slowly* decline (with emphasis on the "slowly" because the dissipation rate is so low). Wouldn't this mean that your great grandchildren wouldn't have to endure the high temperatures that your children will endure? In other words, as the atmospheric CO_2 concentration falls, won't the temperature likewise fall, so that we eventually return to where we started? No. Unfortunately, the response of temperature to changes in the atmospheric CO_2 concentration is not symmetric. If we double the CO_2 concentration the temperature will indeed increase (after a lag of some 20 or 30 years), but if we then cut the CO_2 concentration in half, the temperature will not decrease to what it was before—at least not for several centuries. There is some uncertainty about this, but according to the best current science, over the next century the temperature will go up, but not down—even if CO_2 emissions are cut to zero tomorrow.[4]

(8) **Other Greenhouse Gases and CO_2e:** Other GHGs also contribute to global warming. The most important one is methane (CH_4), discussed below. On a per-ton basis, it is some 28 times as powerful as CO_2 in terms of warming potential. However it dissipates much faster

---

[3] If one ton of CO_2 is emitted today, $(1 - .0035)^{50} = 0.84$ ton will remain after 50 years, and $(1 - .0035)^{100} = 0.70$ ton will remain after 100 years.

[4] This is bad news for your great grandchildren. It is depressing, but does it also seem surprising? See Zickfeld et al. (2013) and Solomon et al. (2009) for detailed discussions.

than CO_2, remaining in the atmosphere only about 8 to 15 years, and this greatly limits its impact. Sometimes methane and the other GHGs are converted to "equivalent" amounts of CO_2 based on their relative warming potential, and the combined measure is called "CO_2-equivalent" (CO_2e). In 2020 the atmospheric CO_2 concentration reached 415 ppm, but the CO_2e concentration was close to 500 ppm. However the use of CO_2e numbers can be misleading because the different GHGs have different dissipation rates and different lag times between changes in concentration and changes in temperature. So I will avoid the use of CO_2e numbers and instead treat the warming effects of CO_2, methane, and other GHGs separately.

(9) **The Importance of CO_2:** Although methane and other GHGs can contribute to climate change, CO_2 is by far the most important problem. One important reason, mentioned above, is that methane and other GHGs simply don't stay in the atmosphere for very long. Most of the methane emitted into the atmosphere this year will be gone in a decade, whereas most of the CO_2 emitted will remain (and cause temperature to rise) for many, many decades. And there is a second reason: the world simply doesn't emit that much methane. On a per-ton basis, annual emissions of methane have been (and continue to be) far less than CO_2 emissions. Later we will see how *rising* methane emissions are problematic, but the overall impact of methane on temperature is much smaller than the impact of CO_2.

Now we can address the question of whether aggressive emission reductions can prevent a temperature increase of 2°C or more by the end of the century.

2.2 An Optimistic Scenario

Later in this book we'll examine in much more detail the temperature implications of alternative paths for CO_2 emissions. We will consider alternative scenarios (some realistic, some less so) for when emissions stop growing and how fast they decline. But for now, let's begin with some simple calculations. We will consider a very optimistic scenario for future global CO_2 emissions—so optimistic that it might be at the limit of what is realistic—and calculate what that scenario would imply for the increase in global mean temperature by the end of the century.

The scenario goes as follows: Although global CO_2 emissions have been steadily increasing (ignoring the drop in 2020 due to the COVID pandemic, at a rate close to 2 percent per year), we will assume that starting in 2020, emissions immediately begin decreasing. From a high of 37 Gt in 2020, we will assume that annual CO_2 emissions decrease along a straight line, reaching zero in 2100. Thus we are assuming that global emissions fall to their 1990 level of about 22 Gt by around 2050, and then continue falling. This path for emissions (from 1960 to 2100) is shown in Figure 2.1. We will ignore emissions of methane, which have been growing sharply and also contribute to warming, and focus just on CO_2.

Aside from the fact that we are ignoring methane, why is this scenario optimistic? After all, the U.S. and Europe are already making headway in emissions reductions. Some states in the U.S. (notably California and New York) have made commitments to reduce emissions sharply over the next 30 years, as have some countries, such as the U.K. Indeed, in a recent article, Heal (2017*b*) has argued that for the U.S., a 50-percent reduction in CO_2 emissions by 2050 is feasible, and although it would be challenging and very costly, an 80-percent reduction by 2050 is possible. (As can be seen in Figure 2.1, our scenario requires only a 40-percent reduction in global emissions by 2050.)

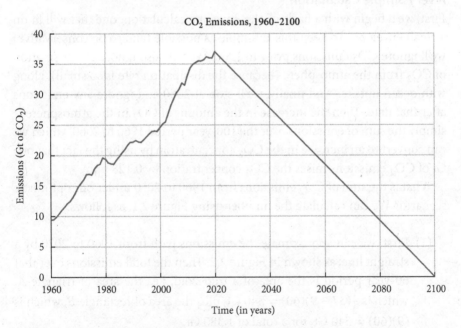

Fig. 2.1 Optimistic scenario—annual global CO_2 emissions decline linearly from 37 Gt in 2020 to zero by 2100. Methane emissions are ignored.

The problem is that what is *feasible* may not actually happen. For example, the U.K. passed a "Climate Change Act" in 2008, which requires GHG emissions in 2050 to be at least 80 percent below their 1990 level, and that target was later made even stricter: net-zero emissions by 2050. But so far the U.K. is not nearly on track to meet even the earlier target. And what happens if we get to 2050 and see that they have missed the target? It is completely unclear (and very unlikely that any of the politicians who devised the target and are still alive will go to jail). Furthermore, what might be feasible for the U.S. and Europe might be much less feasible for China, India, Indonesia, and a range of other countries, where millions of people care much more about escaping dire poverty than they do about protecting the environment.

Look again at Figure 1.2 in Chapter 1. Do you think it is likely that *global* emissions, which has been rising steadily, will suddenly start decreasing and fall steadily to zero by the end of the century? If not, then you'd agree that the scenario is optimistic.

What does this scenario for CO_2 emissions imply for the change in the global mean temperature? We'll address that question with a very simple calculation, and then with a calculation that's a bit less simple.

A Very Simple Calculation

First, we'll begin with a back-of-the-envelope calculation, one that will fit on a small envelope. To keep this as simple as possible (and to be conservative) we'll ignore CO_2 emissions prior to 1960. We will also ignore the dissipation of CO_2 from the atmosphere (because the dissipation rate is so small), along with ocean uptake. And finally, we'll stop at 2060 and ignore any emissions after that date. Then the increase in the amount of CO_2 in the atmosphere is simply the sum of emissions over the 100-year period 1960 to 2060, which we can convert to an increase in the CO_2 concentration by using the fact that one Gt of CO_2 emissions raises the CO_2 concentration by 0.128 ppm.

What is the sum of CO_2 emissions from 1960 to 2060 under our optimistic scenario? We can calculate the number using Figure 2.1, as follows:

(1) First, we can approximate the emissions path from 1960 to 2020 by a straight line, as shown in Figure 2.2. Then the total emissions over that 60-year period is the area of a trapezoid, i.e., the area of triangle A, which is $\frac{1}{2}(37-9)(60) = 840$ Gt, plus the area of rectangle B, which is $(9)(60) = 540$ Gt, for a total of 1,380 Gt.

(2) Next, we want to calculate total emissions over the 40-year period 2020 to 2060. As can be seen in Figure 2.2, total emissions over that period

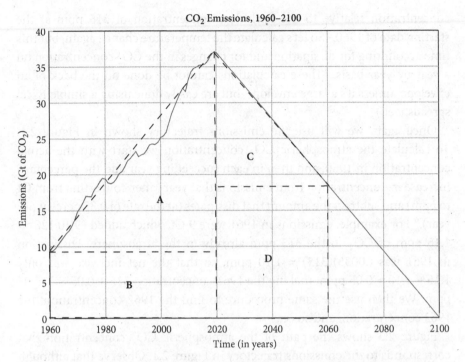

CO₂ Emissions, 1960–2100

Fig. 2.2 Optimistic Scenario—rough calculation of total CO_2 emissions from 1960 through 2060. Total emissions are given by the sum of the areas of triangles A and C and rectangles B and D, which comes to 2,480 Gt. Multiplying by 0.128, this implies an increase in the atmospheric CO_2 concentration of 317 ppm.

is again the area of a trapezoid, in this case the area of triangle C, which is $\frac{1}{2}(37 - 18)(40) = 380$ Gt, plus the area of rectangle D, which is $(18)(40) = 720$ Gt, for a total of 1,100 Gt.

(3) Combining the two periods, total emissions over the 100 years comes to 1,380 + 1,100 = 2,480 Gt. Multiplying by 0.128, this implies an increase in the atmospheric CO_2 concentration of 317 ppm.

The atmospheric CO_2 concentration in 1960 was 315 ppm, so the additional 317 ppm represents just over a 100 percent increase. Using a value of 3.0 for climate sensitivity, this in turn implies an increase in temperature of about 3°C. Yes, that's well above the 2°C limit.

A Less Simple Calculation
You might think that calculation was too simple. After all, we ignored dissipation, and we calculated the percentage increase in the atmospheric CO_2

concentration relative to the fairly low concentration of 315 ppm at the starting date of 1960.[5] So let's calculate the temperature change again, but this time accounting for dissipation and for changes in the CO_2 concentration on a year-by-year basis. (These calculations cannot be done on the back of an envelope, unless it's a large envelope, but are easily done using a simple Excel spreadsheet.)

Once again we will use the emissions trajectory shown in Figure 2.1. To calculate the atmospheric CO_2 concentration we start with the actual concentration in 1960, and then in each succeeding year add the percentage increase in concentration from emissions that year (after converting from Gt to ppm) and subtract the amount that dissipates (at the rate of 0.35 percent per year).[6] For example, emissions in 1961 were 9 Gt, which added $(9)(0.128) =$ 1.15 ppm of CO_2 to the 315 ppm already in the atmosphere. Dissipation in 1961 was $(.0035)(315) = 1.10$ ppm, so that the net increase was only $1.15 - 1.10 = 0.05$ ppm, making the 1961 concentration $315 + 0.05 = 315.05$ ppm. We then use the same procedure to find the 1962 concentration, the 1963 concentration, and so on.

Figure 2.3 shows the path of the atmospheric CO_2 concentration that corresponds to the emissions trajectory in Figure 2.1. Observe that although the concentration initially increases very slowly, the rate of increase rises considerably as emissions rise, so that by 2000 the concentration is 360 ppm. But after 2070 the CO_2 concentration is declining, even though CO_2 emissions are still positive. The reason is that emissions are now low enough, and the concentration high enough, so that the dissipation of the existing stock of CO_2 outweighs the additions to the stock from new emissions.

What would this scenario imply for the increase in global mean temperature? To keep this simple, we will ignore the time it takes for an increase in the atmospheric CO_2 concentration to affect temperature, and assume that any increase in the CO_2 concentration will have an immediate effect on temperature. In our scenario, emissions are assumed to be declining from 2020 onwards, but they are not zero, so the atmospheric CO_2 concentration will still be rising (until 2070, as Figure 2.3 shows). The rising CO_2 concentration will continue to affect temperature, but recall from page 27 that the decline in the concentration will not cause the temperature to go down, at least not by the end of the century.

---

[5] Too simple? Maybe not. In a recent paper Cline (2020) shows that cumulative emissions of CO_2 can provide a reasonable estimate of global warming.

[6] So if E_t is emissions in year t, M_t is the CO_2 concentration, and δ is the dissipation rate, the concentration is given by: $M_t = (1 - \delta)M_{t-1} + E_t$.

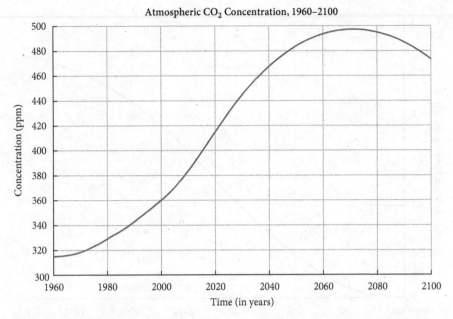

Fig. 2.3 Atmospheric CO_2 concentration under optimistic scenario. By 2070 the concentration reaches a peak (of about 500 ppm) and then starts to decline because dissipation of the stock of CO_2 outweighs additions to the stock from new emissions.

Figure 2.4 shows the cumulative change in temperature relative to 1960 resulting from the change in the atmospheric CO_2 concentration under our optimistic scenario, using a value of 3.0 for climate sensitivity, and assuming that any increase in the CO_2 concentration affects temperature immediately. To calculate the change in temperature, we take the percentage increase in the CO_2 concentration each year and multiply by 3.0 to determine its impact on temperature in the following year.[7] In the year 2000, for example, the CO_2 concentration increased by about 0.5 percent, which results in a $(.005)(3.0) = 0.015°C$ increase in temperature in 2001.

Observe from Figure 2.4 that under this optimistic scenario, the temperature steadily increases, passes the 2°C mark just after 2050, and reaches roughly 2.7°C by 2100. If the correct value for climate sensitivity is indeed 3.0, the rising atmospheric CO_2 concentration, even though rising at a decreasing rate (and falling after 2070), is simply inconsistent with preventing a temperature increase greater than 2°C.

---

[7] Let M_t be the CO_2 concentration in year t, so the percentage increase in the concentration is $(M_t/M_{t-1}) - 1$, which we denote by gM_t. Then the impact of that increase on temperature in the year $t+1$ is $(3.0)gM_t$.

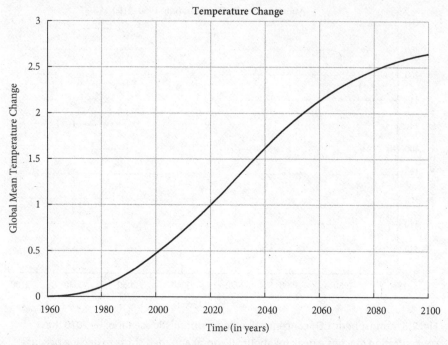

Fig. 2.4 Change in temperature resulting from the change in the atmospheric CO_2 concentration under optimistic scenario. The temperature increases reaches 2°C just after 2050, and reaches about 2.7°C by the end of the century.

2.3 The Bottom Line

What do these rough calculations tell us? Despite efforts to reduce GHG emissions on the part of the U.S., Europe, and some other countries, and despite "commitments" to make further reductions, it is likely that the global mean temperature will increase by more than 2°C, and that might happen as early as 2040. Furthermore, the temperature is likely to continue to increase, reaching something around 3°C by the end of the century. This is under a scenario which, as I explained earlier, is optimistic, because it is very unlikely that *global* emissions, which have been rising steadily, will suddenly start decreasing and fall to zero by the end of the century. In addition, our rough calculations ignored other GHGs, notably methane. Global methane emissions have also been rising, for a number of different reasons. Methane, while it dissipates from the atmosphere much more rapidly than CO_2, has substantial warming potential. (I will discuss methane in more detail later in this book.)

Note that I said "it is *likely* that the global mean temperature will increase by...." It is important to stress the word "likely." As I have said and will explain in detail in the next chapter, there is a great deal of uncertainty when it comes to climate change. Take, for example, climate sensitivity, the increase in temperature resulting from a doubling of the atmospheric CO_2 concentration, once the climate system has come back into equilibrium. The calculations leading to Figure 2.4 were based on a value of 3.0 for climate sensitivity, which is roughly in the middle of the "most likely" range of 1.5 to 4.5, and is currently the IPCC's best estimate. But if we had used the lower end of this range (1.5), the temperature increase by the end of the century would have stayed well below 2°C. And if we had use the higher end of the range (4.5), the calculated temperature increase would have exceeded 4°C. We simply don't know the true value of climate sensitivity (along with a variety of other aspects of the climate system), which means that we can't really say that our scenario implies a temperature increase close to 3°C by the end of the century. All we can say is that a temperature increase of that size or higher is in the realm of what's "likely," or at least quite possible.

But while we can't pin down the temperature increase that would result from our optimistic scenario, the fact that 3°C is likely, or quite possible, or even somewhat possible, still says a lot. It means that despite our best efforts to reduce emissions, we have to be ready for a temperature increase of this magnitude, or even greater, and the effects of such an increase on other aspects of climate. It's nice to be optimistic, but as a matter of public policy, it would be irresponsible to make believe that GHG emission reductions will be sufficient to eliminate the risk of a climate catastrophe.

2.4 Further Readings

Again, this book is not intended to be a thorough introduction to the science and/or economics of climate change. Although I explain in the next chapter what we know and don't know about climate change, my discussion is brief, and some readers might like a more detailed introduction to the subject. For those readers, I would suggest the following books and articles:

- For a thorough and detailed treatment of what we know and don't know about climate change, its impacts, and possible mitigation strategies, at least as of 2014, see the three-volume report of the Intergovernmental Panel on Climate Change, the Panel's 2018 Special Report on the possible impact of a temperature increase above 1.5°C, and the 2021 report on the underlying physical science: Intergovernmental Panel on Climate Change (2014, 2018, 2021).

- The book *Paying for Pollution: Why a Carbon Tax is Good for America* by Metcalf (2019) provides an excellent explanation of why a carbon tax is the most efficient way to reduce CO_2 emissions. But I also recommend this book as an introduction to the economics of climate change.
- For a couple of other articles that are a bit more technical, but still provide an overview of the economics of climate change, see Heal (2017a) and Hsiang and Kopp (2018).
- How difficult would it be to reduce CO_2 emissions by 50 or even 80 percent by mid-century? Heal (2017b) shows how this could be done for the U.S., and explains why a 50-percent reduction in emissions be 2050 is feasible, and although it would be challenging and costly, an 80-percent reduction by 2050 is possible. (But again, this analysis is for U.S. emissions, not global emissions.)
- Why are we waiting (to do something about climate change)? We shouldn't be, and the recent book by Stern (2015) provides a good wake-up call, as well as a nice overview of the science and economics of climate change.
- For a different point of view, Lomborg (2020) argues that climate change, while real and important, is not the kind of emergency that the press and politicians often make it out to be. Lomborg argues that while action on climate change is indeed warranted, many of the policies that have been proposed or adopted will have adverse effects on the economy, poverty, and disease, while doing little to solve the climate problem. In a similar vein, Koonin (2021) stresses the fact that there is much we don't know about climate change. But unlike the message of this book, he ignores the value of insurance and claims that the inherent uncertainty implies that we should wait rather than take action now.

3
What We Know and Don't Know about Climate Change

Before we explore potential solutions to the climate change problem, it's useful to discuss our knowledge of how climate change works in more detail. There is a lot we know about climate change, but there is also a lot we don't know. Even if we knew exactly how much CO_2 and other GHGs the world will emit over the coming decades, we wouldn't be able to predict (with any reasonable precision) how much the global mean temperature will rise as a result. And even if we could predict the amount of warming, we can't predict its *impact*, which in the end is what matters. The fact is that we face considerable uncertainty over climate change and its impact, and as we'll see, that uncertainty has crucial implications for policy.

As explained in the previous chapter, the basic mechanism behind climate change is fairly simple. At the risk of being repetitive, when sunlight reaches the Earth's atmosphere, part of its energy is reflected back into space and the rest is absorbed by the planet. In addition, some energy is always radiating away from the (relatively warm) Earth into (relatively cold) space. We call the difference between the energy flowing in and the energy flowing out *radiative forcing*, and if radiative forcing is positive, the Earth will get warmer. Atmospheric CO_2 increases the fraction of sunlight absorbed relative to the fraction reflected into space, thus warming the planet.

As we burn carbon, more and more CO_2 accumulates in the atmosphere, which increases the amount of radiative forcing, causing an increase in temperatures. Higher temperatures cause other changes in the environment and the climate. For example, warming can cause glaciers in the Arctic and Antarctic regions to break apart, along with thermal expansion of sea water, raising sea levels and possibly inundating coastal areas. And higher ocean temperatures contribute more energy to tropical storms and hurricanes, making them more intense and more destructive.

But *how much* will temperatures rise in response to an increase in the atmospheric CO_2 concentration? And to what extent will higher temperatures cause sea levels to rise and hurricanes to become more intense and

Climate Future: Averting and Adapting to Climate Change. Robert S. Pindyck, Oxford University Press.
© Oxford University Press 2022. DOI: 10.1093/oso/9780197647349.003.0003

destructive? And how will these effects differ across different regions of the world? Unfortunately, we don't know the answers to these questions. That's not to say we don't know anything about the climate effects of a rising CO_2 concentration; we have a reasonable sense of the range of possible outcomes. But that range is fairly wide, and an objective of this chapter is to explain why.

As I said in the Introduction, few would disagree that climate change is on net a bad thing. For the world as a whole, climate change is going to be costly, which is why we need to do something about it. But again, we are faced with considerable uncertainty. Even if we could precisely predict the extent of climate change that we will experience during the coming decades, we wouldn't know what its economic and social impact is likely to be. As we will see, our uncertainty over the cost of climate change has important implications for climate policy.

And there is still another problem. Suppose we could accurately predict not only the extent of climate change, but also its economic impact (expressed in dollars or a percentage reduction in GDP). Most of that economic impact will occur in the distant future, perhaps in 2050 or later. The costs of climate policy, however, will occur much sooner (or at least that's the hope). So, to evaluate any climate policy we need to compare economic impacts in the distant future with policy costs in the near future. That means we need a *discount rate*, so that we can determine the *present value* of these future impacts and costs. And the discount rate we use is very important: A high discount rate implies that today's value of economic impacts occurring in the distant future is low, so there is less need to immediately adopt a stringent emission reduction policy; a low discount rate implies just the opposite. OK, so what is the "correct" discount rate that we should use to evaluate alternative climate policies? It turns out that there is considerable disagreement (at least among economists) as to what the "correct" discount rate is. And yes, this further complicates climate policy.

3.1 The Social Cost of Carbon

To better understand the problem, suppose we did know exactly how much climate change would result from ongoing CO_2 emissions, and suppose we also knew just how costly that climate change will be. In particular, suppose we could determine the impact on the climate of emitting *one additional ton of CO_2 into the atmosphere* today. And suppose we could also determine the future cost of that climate impact, in terms of dollars of lost GDP. And finally, suppose we could agree on a discount rate so that we could determine the present value of that cost, i.e., express it in today's dollars. The cost of that one additional ton of CO_2 is referred to as the *Social Cost of Carbon* (SCC). It is called a "social cost" because the households or firms that emit the CO_2 don't bear the cost of the emissions; instead society does. The cost of emitting the CO_2 is therefore external to the households or firms, which is why we refer to it as an *externality*.[1]

The SCC is the basis for a carbon tax. Why? Because imposing a tax based on the SCC would correct for the fact that households and firms don't bear the full cost of their CO_2 emissions. If you emit a ton of CO_2 and that results in a cost to society of, say, $100, then you should be asked to pay that cost. A carbon tax of (in this example) $100 per ton would correct the problem— you would have to pay $100 for the damage your ton of CO_2 emissions has caused. So, if we could somehow estimate the SCC (on a global basis), we would know how large a carbon tax is needed, and at least in principle we would know how to fix the climate problem. I say "in principle," because a large number of countries would have to agree to this "fix," rather than free ride and let other countries deal with the problem, and that is a tall order. But at least we'd know what is needed.

So how large is the SCC? We don't know. Some argue that climate change will be moderate, will have only a small economic impact on most countries, and will occur in the distant future (making the present value of any economic impact small). This would imply that the SCC is small, perhaps only around $20 per ton of CO_2. Others argue that without an immediate and stringent GHG abatement policy, there is a strong likelihood of substantial temperature increases that might have a catastrophic economic impact, and that impact will occur sooner rather than later. This would suggest that the SCC is large, perhaps $200 per ton of CO_2, or even higher.[2]

Why can't we pinpoint the size of the SCC? Can't we determine the SCC through the use of *integrated assessment models* (IAMs), i.e., models that "integrate" a description of GHG emissions and their impact on temperature (a climate science model) with projections of abatement costs and a description of how changes in climate affect output, consumption, and other economic variables (an economic model)? Building such models is indeed

---

[1] Some microeconomics textbooks, e.g., Pindyck and Rubinfeld (2018), define the social cost of an activity as the *total* private plus external cost. In the climate change literature, however, the term social cost usually refers to the external cost alone, so I will use that definition here.

[2] The highest published estimate I've seen is $400 per ton, in Heal (2020). I recently surveyed several hundred experts in climate science and climate economics to get their opinions on the SCC. I found considerable heterogeneity across experts, leading to a wide variation in the implied SCC numbers. This may simply reflect our very limited knowledge of the underlying science and economics, but it means that there is no single SCC estimate that can be inferred from my survey results. See Pindyck (2019) for details.

what some economists interested in climate change and climate policy have done.[3] The problem is that many of the relationships in these models are ad hoc, with little connection to theory or data, so that the models are not very useful as a policy tool or means of reliably estimating the SCC. (I will discuss the models in more detail later.)

The fundamental problem is that our knowledge of climate change, and especially the economic impact of climate change, is limited. There are some parts of the process that we understand quite well. There is still uncertainty and we might argue about some of the numbers, but at least we have a good idea of what is happening. But then there are other parts that we understand much less well, and parts that we barely understand at all. Here is a brief summary of the mechanisms behind climate change and its impact, with an emphasis on what we know and don't know, and why we don't know certain things.

3.2 Climate Change Basics

To keep things simple, let's ignore methane and other non-CO_2 GHGs for now, and focus on CO_2, which is by far the greatest driver of climate change. To get a better sense of our knowledge and lack thereof, it will be useful to go over the basic mechanisms by which CO_2 emissions originate and accumulate in the atmosphere, how increases in the atmospheric CO_2 concentration leads to climate change, how climate change in turn leads to impacts, and how those impacts can be evaluated in economic terms. We also want to know how emissions can be reduced, and at what cost. We could think about this in terms of a projection of climate damages over the coming century under "business as usual" (BAU), in which nothing is done to reduce emissions, and under alternative emission reduction policies. If we wanted to make such a projection, the steps would be as follows:

(1) **GDP Growth:** GHG emissions are generated by economic activity. If all economic activity throughout the world stopped—no production, no consumption—emissions caused by humans would likewise stop. So, the first step in projecting GHG emissions is to project GDP

---

[3] One of the first such models, developed by William Nordhaus nearly 30 years ago (see Nordhaus (1991, 1993)), was an early attempt to integrate the climate science and economic aspects of the impact of GHG emissions. The early models helped economists understand the basic mechanisms involved by elucidating the dynamic relationships among key variables, and the implications of those relationships, in a coherent and convincing way. But over the past decade or two the models have become large and more complex, but have done little to help us better understand how GHG emissions lead to higher temperatures, which in turn cause (quantifiable) economic damages. My critique of IAMs is summarized in Pindyck (2013a).

growth over the coming century. Not easy! Projecting GDP growth for different countries or regions over the next five years is hard enough. (For example, no one anticipated the worldwide recession that followed the financial crisis of 2008, or the sharp recession caused by the COVID-19 pandemic in 2020.) And now ask, will GDP growth in China, which has been strong up until the pandemic, slow down sharply, or will it pick up again? Will GDP growth in Europe and Japan, which has been anemic over the past decade and has been battered by the COVID-19 pandemic, pick up in the coming several years? We don't know. But even if we could project GDP growth over the next five or ten years it wouldn't get us very far. We need to project GDP growth over at least the next 50 years. Tough job, and clearly subject to considerable uncertainty.

(2) **GHG Emissions:** Marching ahead, let's assume we have a reasonable projection of GDP growth (by region) through the end of the century. We would use this information to make projections of future CO_2 emissions (as well as emissions of other GHGs) under "business as usual" (BAU), i.e., no emission reduction policy in place. To do this, we might relate CO_2 emissions to GDP and then use our projections of future GDP. But this is problematic, in part because the relationship between CO_2 emissions and GDP has been changing, and is likely to continue to change in ways that are not entirely predictable. (The impact of the COVID-19 pandemic, which sharply reduced travel, is an example of how the relationship between CO_2 emissions and GDP can change suddenly and unpredictably.) We would face the same problem when we try to project CO_2 emissions under one or more abatement scenarios, or under alternative scenarios for GDP growth.

(3) **Atmospheric GHG Concentrations:** Suppose we have projections of CO_2 emissions through the end of the century. We could use those projections to in turn project future atmospheric CO_2 concentrations, accounting for past and current emissions as well as future emissions. (We could do the same thing for methane, but since it dissipates from the atmosphere relatively rapidly, we'll focus on CO_2.) There is some uncertainty here, because the CO_2 dissipation rate depends in part on the total concentrations of CO_2 in the atmosphere and in the oceans. But relative to other uncertainties (discussed below), translating emissions to concentrations can be done with reasonable accuracy.

(4) **Temperature Change:** Now we come to the hard part. We would like to make projections of average global (or regional) temperature changes—and possibly other measures of climate change such as

rainfall variability, hurricane frequency and intensity, and sea level increases—that are likely to result from higher CO_2 concentrations. Can't we project the temperature change likely to result from any particular increase in the CO_2 concentration by applying a value for climate sensitivity? In Chapter 2 we did just that (for an "optimistic" CO_2 emission scenario), using a mid-range value of 3.0 for climate sensitivity. But as I explained, we don't know the true value of climate sensitivity. Until 2021, the IPCC declared the "most likely" range to be from 1.5 to 4.5, which in 2021 they narrowed to be from 2.5 to 4.0. If we include what the IPCC has considered "less likely" but possible values, the range would run from 1.0 to 6.0. Even 2.5 to 4.0 is a large range, and implies a large range for temperature change. On top of that uncertainty, what is the time lag between an increase in the CO_2 concentration and its impact on temperature? Something like 10 to 50 years, but again, that is a wide range.

(5) **Impact of Climate Change:** But let's keep marching ahead and assume that we know how much the temperature will increase during the coming decades (and how much sea levels will rise, etc.), and try to project the economic impact of such changes in terms of lost GDP and consumption. Now we are in truly uncharted territory. Most integrated assessment models (IAMs) make such projections by including a "damage function" that relates temperature change to lost GDP, but those damage functions are not based on any economic (or other) theory, nor on any empirical evidence. They are essentially arbitrary functions, made up to describe how GDP might go down when temperature goes up. To make matters worse, "economic impact" should include indirect impacts, such as the social, political, and health impacts of climate change, which might somehow be monetized and added to lost GDP. Here, too, we are in the dark. There is speculation about the possible social and political impacts of higher temperatures, and how higher temperatures might affect mortality and morbidity rates in different countries, but there is little in the way of actual empirical evidence. Basically, we just don't know what the true damage function looks like. The bottom line: Projecting the impact of climate change is the most speculative part of the analysis.

(6) **Abatement Costs:** To evaluate a candidate climate policy, we must compare the benefits of the policy to its costs. What are the benefits? A reduction in climate-induced damages, e.g., a reduction in the loss of GDP that would otherwise result from climate change. Estimating these benefits is a problem because, as explained above, projecting the impact of climate change is highly speculative. And what about the

costs of a candidate climate policy? Determining the costs requires estimates of the cost of abating GHG emissions by various amounts, both now and throughout the future. A small amount of abatement (say, reducing CO_2 emissions by 10 percent) is fairly easy, but a large amount (say, cutting emissions by 70 or 80 percent) is likely to be very costly. But how costly? We're not sure, in part because we have had no experience cutting emissions by 70 percent or more. And how will abatement costs change over the coming decades? Answering that question requires projections of technological change that might reduce future abatement costs, and technological change is very hard to predict. Once again, we face considerable uncertainty.

(7) **Valuing Current and Future Losses of GDP:** Finally, let's assume that we could somehow determine the annual economic losses (measured in terms of lost GDP) resulting from any particular increase in temperature. Let's also assume that we know the increases in temperature that would result from "business as usual" (BAU) and, alternatively, under some particular emission abatement policy. And suppose we also know the annual cost (again in terms of lost GDP) of that abatement policy. How would we evaluate the policy? In other words, how would we compare the benefits from the policy to its costs? We would need to know the *discount rate* that would let us compare current losses of consumption (from the cost of abating emissions) with the future gains in consumption from the reduction in economic losses resulting from the abatement policy.[4] The discount rate (in this case the *social rate of time preference*, because it measures how society values a loss of consumption in the future versus today) is critical: A low discount rate (say around 1 percent) makes it easier to justify the immediate adoption of a stringent abatement policy; a high rate (say around 5 percent) does the opposite. So, what is the "correct" discount rate? As we'll see, there is no clear number on which economists agree.

To summarize, there are aspects of climate change, in particular GHG emissions and concentrations, where we have a reasonable amount of knowledge and can make reasonable projections. Yes, there is uncertainty, especially when projecting out 50 or more years. But at least we can pinpoint the nature of the uncertainty, and to some extent bound it. And then there are aspects of climate change—changes in temperature, sea levels, and hurricane

---

[4] We might also want to know the *social welfare function*, i.e., the loss of social utility resulting from a loss of GDP (and hence from a loss of consumption). If GDP and consumption are very high, the loss of utility resulting from a 5 percent loss of GDP would be smaller than if GDP and consumption were low. I will discuss this point later.

intensity, and most notably, the economic impact of those changes—where we know very little. Let's turn now to a more detailed discussion of what we know and don't know, and try to understand *why* we don't know certain things, the extent of the uncertainty, and the likelihood that the uncertainty will be reduced during the coming years.

3.3 What We Know (or Sort of Know)

There are some parts of the climate change process that we understand fairly well. There is still considerable uncertainty regarding the specific numbers, but at least we can estimate those numbers and come up with reasonable bounds.

3.3.1 What Drives CO_2 Emissions?

Burning carbon creates CO_2 emissions. That's easy, but just how much carbon will be burned over the coming decades, and how much CO_2 will be emitted? The answer depends on economic activity and efforts at emissions abatement. Let's put aside emissions abatement for now, because we'll deal with that when we discuss climate change policy. What drives CO_2 emissions absent any abatement policy? Economic activity.

Much of economic activity—whether the production or the consumption of goods and services—involves the burning of carbon. On the production side, factories require energy to operate, and much of that energy comes from fossil fuels, either directly (e.g., burning coal and residual fuel oil to produce iron, steel, copper and other metals) or indirectly (e.g., burning fossil fuels to produce electricity, which in turn is used to produce aluminum). On the consumption side, we burn fossil fuels to heat our homes, drive our cars, and fly around the world. Thus as economic activity, measured, say, by GDP, grows, CO_2 emissions will grow as well.

Suppose we have a projection for GDP growth over the next several decades. Can we use that to project CO_2 emissions? It's a bit complicated, as we'll see.

Carbon Intensity

The relationship between GDP and CO_2 emissions is neither simple nor fixed. Over the past 50 years or so, the amount of CO_2 emitted per dollar of GDP has declined steadily—in the U.S., in Europe, in China, and in most other countries. This ratio, CO_2 emitted per dollar of GDP, is called *carbon intensity*. Carbon intensity has been declining for several reasons:

- The composition of GDP, i.e., the mix of goods and services that makes up GDP, has been changing. Compared to 50 years ago, services (e.g., medical care, entertainment, retailing, etc.) has become relatively more important than manufacturing and transportation, and services use less energy and therefore emit less CO_2 than manufacturing and transportation.
- Technological improvements in the way we produce and utilize goods and services has resulted in the use of less energy, and thus lower emissions of CO_2. For example, cars, trucks and buses are much more fuel efficient than they were 50 years ago, as are home and commercial heating and cooling systems.
- Energy itself is becoming "greener." Energy generation from renewables (especially wind and solar) has been growing, and the share of energy coming from fossil fuels, especially coal, has been falling. Also, the shift from coal to natural gas, while not "green," cuts CO_2 emissions in half.

To understand what might happen in the future, let's take this a step further by breaking carbon intensity down into its components, which can be measured and understood as follows:

(1) **Energy Intensity:** The amount of energy consumed per dollar of GDP. We measure energy consumption in quadrillions of BTUs (10^{15} BTUs, denoted as *quads*), and GDP in billions of dollars, usually U.S. dollars.[5] For international comparisons, we convert a country's GDP into U.S. dollars using either an exchange rate or a Purchasing Power Parity Index.[6] So the unit of measurement for energy intensity is quad BTUs/$ billion.

(2) **Energy Efficiency:** Sometimes referred to instead as *carbon efficiency* or CO_2 *efficiency*, this is the amount of CO_2 emitted from the consumption of 1 quad of energy. If, for example, the energy is generated from

[5] One BTU (British thermal unit) is the amount of heat energy required to raise the temperature of one pound of water by one degree Fahrenheit. In the metric system, the unit of energy is the calorie, which is the amount of heat required to raise the temperature of one gram of water by 1°C. One BTU is approximately equal to 252 calories.

[6] Exchange rates are determined by flows of traded goods and flows of capital, but many of the goods people consume are not traded (e.g., housing, transportation, and food) and people do not (directly) consume capital. Unlike an exchange rate, a Purchasing Power Parity (PPP) Index allows us to convert currencies from one country to another in terms of what people can actually consume.

wind or solar, little or no CO_2 will be emitted. A moderate amount of CO_2 is emitted if the energy is generated from natural gas, and a large amount is emitted if the energy is from coal. For energy efficiency, we measure CO_2 emissions in *megatons* (Mt, millions of metric tons), so the unit of measurement for energy efficiency is Mt of CO_2/quad BTUs.

(3) **Carbon Intensity:** The amount of CO_2 emitted (in megatons) per billion dollars of GDP. Carbon intensity is simply the product of energy intensity and energy efficiency:

$$\text{Carbon Intensity} = \text{Mt } CO_2/\$ \text{billion} = (\text{quads}/\$ \text{billion}) \times (\text{Mt } CO_2/\text{quad})$$

Decomposing carbon intensity into its two components is useful because the drivers of energy intensity and energy efficiency can be quite different.

What does this decomposition of carbon intensity tell us? It says that if we want to predict CO_2 emissions over the coming decades (with or without some abatement policy), we would have to do the following:

(1) predict GDP growth;
(2) predict changes in energy intensity; and
(3) predict changes in energy efficiency.

Furthermore, we'd have to do this for every major country, or at least different regions of the world, because GDP growth, energy intensity and energy efficiency are likely to evolve very differently in different countries and regions. This is explained in more detail below.

GDP Growth

Let's begin with GDP growth. Figure 3.1 shows the GDP for the U.S., Japan, China, and India in constant 2010 U.S. dollars, from 1960 to 2018. For the U.S., real (i.e., net of inflation) GDP growth has fluctuated from negative values during recessions (for example, −1.8 percent in 1982 and −2.5 percent in 2009) to large positive values during recoveries (e.g., 7.2 percent in 1984), but has averaged around 2.0 to 2.5 percent per year. Not shown is the very deep recession in 2020 that most of the world has encountered because of the COVID-19 pandemic. Depending on the recovery from this pandemic-induced recession, we might expect a return to historical growth rates for the U.S. over the next one or even two decades. But can we expect to see the same growth rates over the rest of the century? We don't know.

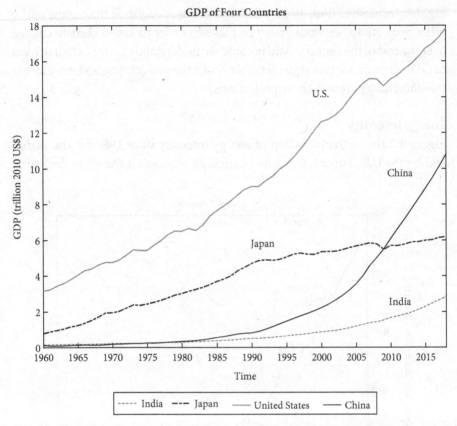

Fig. 3.1 GDP of the U.S., China, Japan, and India, 1960–2018, in 2010 US Dollars.
Source: World Bank.

Projecting GDP growth for the other three countries shown in Figure 3.1 is even more problematic. Japan's GDP grew at about 5 to 6 percent annually from 1960 to 1992, but then grew at less than 1 percent annually from 1992 onwards. Will post-pandemic economic growth in Japan pick up during the next several decades? We don't know. China's experience was the opposite: slow growth during the 1960s, 1970s, and early 1980s, and then an average growth rate of around 9 percent from 1990 onwards. The COVID-19 pandemic caused a sharp decline in China's GDP, but even with a full recovery, will the growth of China's economy slow down over the coming decades? Probably, but by how much? In terms of population, China and India are the largest countries in the world. What will their GDPs look like over the rest of the century?

Predicting GDP growth for any country, even just several years out, is notoriously difficult. Predicting GDP growth over the next 10, 20, or 50 years for different countries with quite different economies is clearly highly speculative.[7] But, as we will see, it is less speculative than some of the other projections we would need to make in order to assess climate change over the rest of the century. And because we understand the important drivers of GDP growth, we can argue sensibly about the projections, and even assess the uncertainty around those projections.

Energy Intensity

Figure 3.2 shows the evolution of energy intensity since 1980 for the world, and for the U.S., Europe, India, and China. (Not shown is the sharp decline in

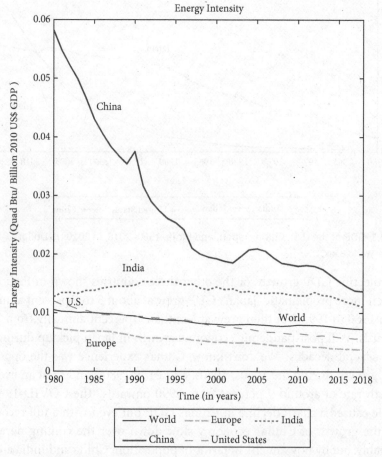

Fig. 3.2 Energy Intensity for the world, and for the U.S., Europe, India, and China. Energy Intensity is measured in quad (10^{15}) BTUs per Billion 2010 U.S. dollars of GDP. *Source*: World Bank, U.S. Energy Information Agency.

---

[7] In a recent study, Müller, Stock, and Watson (2019) pool data for 113 countries over the 118 years from 1900 to 2017 to model the behavior of long-run GDP growth. They find that pooling across countries yields tighter prediction intervals, but even with more than a century of data, "the 100-year growth paths exhibit very wide uncertainty."

energy intensity in 2020 caused by the COVID-19 pandemic, which severely limited travel and thus the consumption of gasoline and jet fuel. That decline may or may not be temporary.) Observe that for the U.S. and Europe, energy intensity has declined steadily (although energy intensity has always been higher in the U.S.). This decline is largely due to gradual changes in the composition of GDP in the U.S. and Europe, and the ways in which GDP is produced and consumed. Compared to 1980, services (e.g., medical care, insurance, retailing) are a larger share of GDP, and for the most part the production of services uses less energy than the production of manufactured goods. In addition, improvements in the way we produce and utilize goods and services has resulted in the use of less energy. For example, cars and trucks have become much more fuel efficient, as have household appliances (such as refrigerators, washing machines, and televisions) as well as home and commercial heating and cooling systems.

As Figure 3.2 makes clear, China has exhibited the largest decline in energy intensity. Energy intensity in China was extremely high in 1980 (about five times the world average), in part because the Chinese GDP was very low, and in part because relatively simple changes in the production and use of goods and services could bring about large declines in energy use. As a result, Chinese energy intensity fell from almost 0.06 quads/$ Billion in 1980 to about .02 in 2000. But since 2000, the rate of decline has been much smaller, to about .015 quads/$ Billion in 2016. Why? In part because manufacturing has been increasing, and Chinese consumers want—and buy—many more cars, household appliances, and travel, all of which require more energy.

Despite the large decline in energy intensity in China, and the declines in the U.S. and Europe, for the world as a whole reductions in energy intensity have been quite limited; a decline from .0110 to .0075 quads/$ Billion. In part, this is because there has been little or no decline in energy intensity for other large developing countries. For example, as Figure 3.2 shows, there has been almost no decline in energy intensity in India. So the question now is whether we should expect energy intensity on a worldwide basis to decline much further, or level out at close to its current value. If energy intensity does not decline significantly, it will be difficult to achieve a decline in carbon intensity.

Energy Efficiency

Even if energy intensity remains constant, we would see a reduction in carbon intensity if we could achieve a significant improvement in energy efficiency. In other words, could we reduce the amount of CO_2 that results from the consumption of each BTU of energy? To answer this question, we can begin

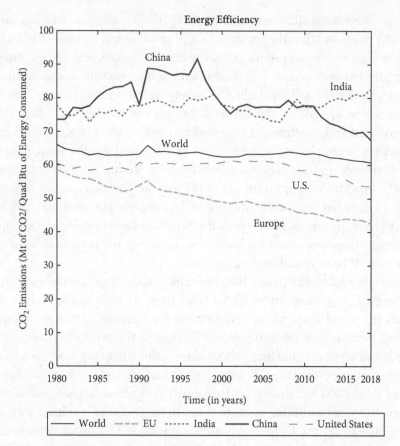

Fig. 3.3 Energy Efficiency for the world, and for the U.S., Europe, India, and China. Energy Efficiency is measured in Mt of CO_2 emissions per quad BTU of energy consumed, so a *reduction* in energy efficiency implies an improvement, i.e., a reduction in the amount of CO_2 generated from the use of energy.
Source: U.S. Energy Information Agency.

by looking at changes in energy efficiency over the past few decades. Figure 3.3 shows the evolution of energy efficiency since 1980 for the world, and for the U.S., Europe, India, and China. (Recall that energy efficiency is measured as Mt of CO_2 emissions per quad BTU of energy consumed.)

Observe that both Europe and (to a lesser extent) the U.S. have had improvements in energy efficiency. In Europe, energy efficiency was about 67 Mt CO_2/quad BTU in 1980, and only about 43 Mt CO_2/quad BTU in 2018. In the U.S. there was almost no change from 1980 to 2005 (about 60 Mt CO_2/quad BTU in both years), but then in the next 10 years there was a decline to about 54 Mt CO_2/quad BTU. These changes have occurred because in Europe and the U.S., energy is becoming "greener." Energy generation from renewables (especially wind and solar) has been growing, and the share of energy coming from fossil fuels, especially coal, has been falling.

But alas, energy efficiency in China and India in 2018 is close to where it was in 1980—nearly 70 Mt CO_2/quad BTU in China and around 82 Mt CO_2/quad BTU in India—and well above the levels in the U.S. and Europe. Energy efficiency has followed a similar pattern in other large developing countries. Why? Because there has been little or no change in the way energy is produced in these countries. Yes, we are beginning to see more use of renewables such as wind and solar, but the amount is still very small, and rising only slowly. The net result: On a worldwide basis, energy efficiency has remained roughly constant (at about 60 Mt CO_2/quad BTU).

So as with energy intensity, the question now is whether we should expect energy efficiency on a worldwide basis to improve significantly, or remain close to its current value. If energy efficiency does not improve, it will be difficult to achieve a decline in carbon intensity. And then, unless economic growth comes to a halt, emissions of CO_2 will keep increasing.

Carbon Intensity

Finally, Figure 3.4 shows the product of energy intensity and energy efficiency, i.e., carbon intensity, measured in Mt of CO_2 emitted per $ Billion of GDP. The graph is similar in shape to Figure 3.2 for energy intensity, and that's because except for Europe and (to a lesser extent) the U.S., there has been little improvement in energy efficiency. So for Europe and the U.S. there has been a gradual decline in carbon intensity, and for China a sharp decline, mirroring the decline in China's energy intensity. And for the world as a whole? A gradual decline in carbon intensity from about 0.69 Mt CO_2/$ Billion in 1980 to 0.50 Mt CO_2/$ Billion in 2000, but after 2000 almost no further decline; it was just under 0.50 Mt CO_2/$ Billion in 2018.

What does a decline in worldwide carbon intensity from 0.69 Mt CO_2/$ Billion in 1980 to 0.50 Mt CO_2/$ Billion in 2018 imply for worldwide CO_2 emissions? Worldwide carbon intensity declined by about 30 percent, so if world GDP had remained constant over that time period, CO_2 emissions would have likewise declined by about 30 percent. But (fortunately) world GDP has grown substantially. Measured in 2010 constant U.S. dollars, it tripled, going from about $28 trillion in 1980 to about $84 trillion in 2018. And that's why global CO_2 emissions have increased so much.

To illustrate this problem further, we can calculate the CO_2 emissions by multiplying world GDP and worldwide carbon intensity. For 1980, carbon intensity was about 0.69 Mt CO_2/$ Billion, which is equivalent to 0.69 Gt CO_2/$ Trillion. World GDP was about $28 Trillion, which implies CO_2

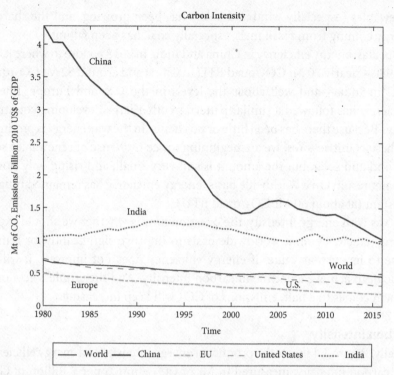

Fig. 3.4 Carbon Intensity for the world, and for the U.S., Europe, India, and China. Carbon Intensity is the product of energy intensity (Fig. 3.2) and energy efficiency (Fig. 3.3), and is measured in Mt of CO_2 emissions per billion 2010 U.S. dollars of GDP.

emissions of $(0.69)(28) = 19$ Gt, very close to its actual value. And for 2018, carbon intensity was 0.50 Mt CO_2/$ Billion = 0.50 Gt CO_2/$ Trillion, and world GDP was about $84 Trillion, which implies CO_2 emissions of $(0.50)(84) = 42$ Gt, somewhat above its actual value.

The bottom line: a modest decline in worldwide carbon intensity combined with a large increase in world GDP has resulted in a near-doubling of global CO_2 emissions.

Future CO_2 Emissions

Where does this leave us in terms of projecting future CO_2 emissions? On one level it paints a rather grim picture. On a worldwide basis, carbon intensity has been declining very slowly, only by about 1 percent per year from 1980 to 2018. World GDP on the other hand, has been growing much faster, at an average rate of about 3 percent per year. So there are only two ways that CO_2 emissions can decline in the future: (1) a decline in world GDP; or (2) a decline in worldwide carbon intensity. A decline in world GDP, or even a reduction in the rate of GDP growth, is not a happy thought. And we certainly wouldn't want to engineer a global recession (or depression) as a means of reducing CO_2 emissions. So that leaves us with the second option—a decline in carbon intensity.

A decline in carbon intensity can result from a decline in energy intensity, and/or a decline (i.e., improvement) in energy efficiency. Do we have any good reasons to expect either to occur? Yes and no.

Yes, because energy intensity and energy efficiency can both be affected by government policy. Indeed, that is what much of climate policy is all about. Think what would happen if the world adopted a carbon tax, which would raise the price of burning carbon. Since most of the energy we use is generated from fossil fuels, i.e., from carbon, such a tax would reduce our use of energy. In other words, the tax would cause a decline in energy intensity. And there are other policy options in addition to a carbon tax, such as automobile fuel efficiency standards, building codes that require more insulation, and related measures that would directly reduce energy intensity. Indeed, technological improvements in the way we produce and utilize goods and services has already resulted in the use of less energy; cars, trucks, and buses are much more fuel efficient than they were 30 years ago, as are home and commercial heating and cooling systems.

A carbon tax would also create incentives to produce energy with less carbon, and thereby improve energy efficiency. For example, a BTU of energy obtained from burning natural gas produces about half as much CO_2 compared to a BTU obtained by burning coal, so the tax would result in a more rapid shift from coal to natural gas. (That shift away from coal could also be achieved via direct regulations over the construction of new power plants.) Likewise, a BTU of energy obtained from wind power burns no carbon, and thus becomes more attractive once a carbon tax is in place.

But I said "yes and no." Why "no?" Because we have to consider the cost of these policies, their political feasibility, and the problem of free-riding by some countries. Start with the cost. What would a carbon tax, fuel efficiency standards, and other policy measures cost in terms of reduced consumption (private and public) and the growth of consumption? We're not sure. Estimates of CO_2 abatement costs *today* vary widely, and future abatement costs are much more uncertain because we can't predict the technological change that might reduce those costs.

Even if the cost of a strong abatement policy is moderate (perhaps reducing personal consumption be just a few percent) would it be politically feasible? Put another way, suppose we use a carbon tax to reduce emissions. Then how large a tax can we reasonably expect to see? Or if a cap-and-trade system is used to reduce emissions, so that permits are required to emit CO_2, how high a permit price can we expect? The answers to these questions will vary greatly across countries. As of this writing, the adoption of a strong abatement policy seems quite likely in Europe, but less so in the U.S., and much less so in key countries such as China, India, Indonesia, and Russia. Closely related to this is the free-riding problem, which reduces the political feasibility of strong abatement policies in many countries.

I will discuss these policy issues in more detail later in this book. At this point, let's summarize where we are regarding future CO_2 emissions: If we could predict the growth of GDP around the world, *and* predict the changes in energy intensity and energy efficiency, and thus the change in carbon intensity, we could come up with at least a rough prediction of future CO_2 emissions. And we would want to make that rough prediction under "business as usual," and under one or more CO_2 abatement policies.

3.3.2 What Drives the Atmospheric CO_2 Concentration?

Even though they will be subject to uncertainty, making predictions of future CO_2 emissions, especially under alternative climate policies, is an important task. But remember that CO_2 emissions do not *directly* cause increases in temperature. Instead, warming is caused by increases in the atmospheric CO_2 *concentration*. Of course, increases in the CO_2 concentration are the result of CO_2 emissions, so if we want to make predictions about increases in temperature, we need to determine how any particular path for emissions affects the future path of the CO_2 concentration.

Isn't the current atmospheric CO_2 concentration just the sum of past emissions, minus any dissipation? Roughly, but not precisely. The problem is that some atmospheric CO_2 is absorbed by the oceans, and some of the CO_2 in the oceans can re-enter the atmosphere. How much goes each way? That depends on a variety of factors, including the amounts of CO_2 both in the atmosphere and in the oceans, and the ocean temperature. So even if we had precise projections of CO_2 emissions over the next several decades, our projection of the atmospheric CO_2 concentration would be subject to some uncertainty.[8] Nonetheless, compared to some of the other

---

[8] A variety of large-scale "General Circulation Models" (GCMs) have been developed to address this problem and provide tighter estimates of the path for the atmospheric CO_2 concentration—and the path for global mean temperature—that will result from a particular path for CO_2 emissions. A related model is the MIT Earth System Model developed as part of the MIT Joint Program on the Science and Policy of Global Change. For examples of the application of that model, see Paltsev et al. (2016) and Sokolov et al. (2017).

uncertainties we face (see below), this one is not too bad. Given a predicted path for CO_2 emissions, we can predict the atmospheric CO_2 concentration reasonably well.

At the most basic level, if we ignore movements of CO_2 into and from the oceans, adding up past CO_2 emissions and subtracting dissipation will give us reasonable estimate of the atmospheric concentration. We did this in Chapter 2, when we calculated the temperature implications of a scenario in which CO_2 emissions stop increasing in 2020, and decline linearly to zero by 2100, as shown in Figure 2.1. We first had to calculate the resulting path for the atmospheric CO_2 concentration. To do that we began with the actual concentration in 1960, and then in each succeeding year added the increase in concentration from that year's emissions (after converting from Gt to ppm) and subtracted the amount that dissipates (at the rate of 0.35 percent per year).[9]

For example, emissions in 1961 were 9 Gt, which added (9)(0.128) = 1.15 ppm of CO_2 to the 315 ppm already in the atmosphere. Dissipation in 1961 was (.0035)(315) = 1.10 ppm, so the net increase was 1.15 − 1.10 = 0.05 ppm, making the 1961 concentration 315 + 0.05 = 315.05 ppm. In this way we calculated the resulting path for the atmospheric CO_2 concentration, which is shown in Figure 2.3. This calculation is far from perfect, but it provides a reasonable first cut at how CO_2 emissions lead to changes in the CO_2 concentration.

In Chapter 2 we went further and estimated the impact on temperature of this path for the CO_2 concentration. That required a value for climate sensitivity, which connects changes in the CO_2 concentration to changes in temperature. It is at this point where the uncertainty becomes greater—much greater—as we'll see below.

3.4 What We Don't Know

Now we come to the hard part. We would like to make projections of average global (or regional) temperature changes, and other measures of climate change such as rainfall variability, hurricane frequency and intensity, and sea level increases, that are likely to result from higher CO_2 concentrations. We

---

[9] We saw earlier that ignoring movements of CO_2 into and from the oceans, we could write the relationship between CO_2 emissions and the CO_2 concentration as follows: Denoting emissions in year t by E_t, the concentration by M_t, and the dissipation rate by δ, the concentration is given by:

$$M_t = (1 - \delta)M_{t-1} + E_t.$$

can make such projections, but they will be subject to extreme uncertainty. And then given those projections of climate change, we would like know their probable *impact*, i.e., the extent to which climate change would result in lost GDP and consumption, as well as higher morbidity and mortality, and other measures of damages. Here we are in uncharted territory.

Can't we project the temperature change likely to result from any particular increase in the CO_2 concentration by applying a value for climate sensitivity? In Chapter 2, we did just that (for an "optimistic" CO_2 emission scenario), using a mid-range value (3.0) for climate sensitivity. But as I explained, we don't know the *true value* of climate sensitivity. According to the most recent (2021) report, the "most likely" range is from 2.5 to 4.0, and if we include what the IPCC has considered "less likely" but possible values, the range would run from 1.0 to 6.0. Why can't we narrow down the range of possible values for climate sensitivity? Is it likely that our understanding of climate science will improve over the coming decade, in a way that will reduce the uncertainty. I address these questions below.

Even if we knew how much the temperature will increase during the coming decades (and how much sea levels will rise, etc.), what matters is the likely impact of those changes. If higher temperatures and higher sea levels cause little damage, why should we devote resources today on preventative measures? On the other hand, if the likely damages are extreme, then we certainly should act quickly to reduce emissions and prevent climate change. Thus it is important to determine the likely economic impact of warming, rising sea levels, and other measures of climate change in terms of lost GDP and consumption. Furthermore, "economic impact" should include indirect impacts, such as the social, political, and health impacts of climate change, which we might try to monetize and include in a broader measure of lost GDP. Unfortunately, when it comes to the impact of climate change we are in the dark, and can only speculate.

Why is it so difficult to pinpoint climate sensitivity, or at least narrow the range of estimates? Why can't we predict the likely impact of climate change on the economy? We now turn to these questions.

3.4.1 Climate Sensitivity

Recall that climate sensitivity is defined as the temperature increase that would eventually result from an anthropomorphic doubling of the atmospheric CO_2 concentration. The word "eventually" means after the world's climate system reaches a new equilibrium following the doubling of the CO_2

concentration. It would take a very long time, however, for the climate system to *completely* reach a new equilibrium, around 300 years or more. However, the climate system will get quite close to equilibrium in a few decades. How many decades depends in part on the size of the increase in the CO_2 concentration—the larger the increase, the longer is the time lag—and even for a given increase, there is some uncertainty over the time lag. But in most cases, 10 to 40 years is a reasonable range, and 20 or 30 years are commonly used numbers.[10]

I said that there is considerable uncertainty over the true value of climate sensitivity. Three questions come up:

(1) First, just how much uncertainty is there? To what extent can we narrow the range of possible values?

(2) Second, there has been a good deal of research in climate science during the past few decades. Has that research led to a better understanding of the mechanisms underlying climate sensitivity, in a way that would allow us to obtain more precise estimates? In other words, has the uncertainty over climate sensitivity been reduced, and if so, by how much?

(3) Finally, *why* is there so much uncertainty? What is it about climate sensitivity that prevents us from obtaining precise estimates of its value?

Let's address each of these questions in turn.

How Much Uncertainty Is There?

Over the past two decades there have been a large number of studies by climate scientists on the magnitude of climate sensitivity. Virtually all of those studies conclude by providing a range of estimates, often in the form of a probability distribution. From the probability distribution we can determine the probability that the true value of climate sensitivity is above or below any particular value, or within any interval; for example, above 4.0°C, or between 2.0 and 3.0°C. Thus, each study gives us an estimate of the nature and extent of uncertainty, *according to that study*. One example of such a study

---

[10] Climate scientists often distinguish between "equilibrium climate sensitivity," which is climate sensitivity as I have described it above, and "transient climate response," which is the response of global mean temperature to a gradual (1-percent per year) increase in the CO_2 concentration. See National Academy of Sciences (2017), pages 88–95, for a discussion. I will simply use the term "climate sensitivity," and treat the time lag (10 to 40 years) as the time it takes for the climate system to get close to equilibrium.

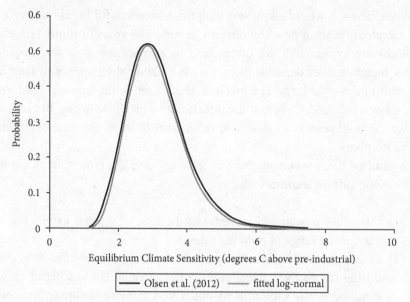

Fig. 3.5 Probability distribution for climate sensitivity (original and log-normal fit) from Olsen et al. (2012). This figure is from the Web Appendix of Gillingham et al. (2018), and used by permission.

is Olsen et al. (2012). The probability distribution arrived at by that study is shown in Figure 3.5.[11]

The mean value of climate sensitivity according to this probability distribution is 3.1°C. To determine the probability that the true value of climate sensitivity is within any particular range, we just have to find the corresponding area under the curve in Figure 3.5. For example, the area under the curve between 1.8 and 4.9 is 0.95, which implies that there is a 95 percent probability that the true value of climate sensitivity is between 1.8 and 4.9°C. (The authors of the study call this the "credible interval.") One might also conclude from this probability distribution that it is likely (i.e., the probability is above 0.75) that the true value is between 2.0 and 4.0.

Why doesn't the Olsen et al. (2012) study give us a precise number for climate sensitivity, rather than the distribution shown in Figure 3.5? Because the authors of the study used a model of the climate system and recognized that some of the parameters of the model are uncertain. They attached probability distributions to those parameters, and thereby obtained a probability distribution for the model's projection of climate sensitivity.

---

[11] This figure is taken from the Web Appendix of Gillingham et al. (2018), and used by permission. The figure includes the original distribution obtained by Olsen et al. (2012) as well as a log-normal distribution fitted to the original one. (The two are almost overlapping.)

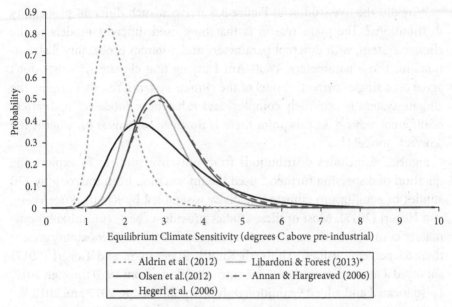

Fig. 3.6 Log-normal distributions fit to the probability density functions cited in IPCC, 2014. This figure is from the Web Appendix of Gillingham et al. (2018), and is used by permission.

But now we have to be careful. Drawing conclusions about climate sensitivity from Figure 3.5 could be misleading, and in particular could lead us to underestimate the extent of the uncertainty. The problem is that Olsen et al. (2012) is only one study. There have been many other studies, and those different studies, which utilize different climate models, arrive at different probability distributions. In other words, the studies disagree considerably on both the extent of the uncertainty, and its characteristics (such as the mean value). This suggests that the actual uncertainty is much greater than what is indicated by any single study.

The dispersion across studies is illustrated by Figure 3.6, which shows the probability distributions from Olsen et al. (2012) and four other studies.[12] Observe that these five distributions differ considerably. One of them, Aldrin et al. (2012), shows a relatively narrow and low range for climate sensitivity, and implies that there is a 95 percent probability that the true value is between 0.8 and 3.0°C. Another, Hegerl et al. (2006), shows a relatively wide range for climate sensitivity, from 1.0 to about 6.0°C.

[12] The other studies are Aldrin et al. (2012), Libardoni and Forest (2013), Annan and Hargreaves (2006), and Hegerl et al. (2006). The figure shows log-normal distributions that were fitted to the original distributions in the five studies. It is taken from the Web Appendix of Gillingham et al. (2018), and used by permission.

Why did the five studies in Figure 3.6 arrive at such different probability distributions? The main reason is that they used different models of the climate system, with different parameters and different probability distributions for those parameters. Wait! Am I saying that climate scientists don't agree on a single "correct" model of the climate system? Yes, that's right. The climate system is extremely complex, and it has been modeled in a variety of different ways.[13] At this point there is no clear consensus on what is the "correct" model.[14]

Figure 3.6 includes distributions from only five studies. To explore the question of dispersion further, I used the information from the roughly 130 studies of equilibrium climate sensitivity assembled by Knutti, Rugenstein, and Hegerl (2017). Most of these studies provide a "best" (most likely) estimate of climate sensitivity, as well as a range of "likely" (i.e., probability greater than 66 percent) values. Although Knutti, Rugenstein, and Hegerl (2017) surveyed a few earlier studies, I only included those from 1970 through 2017. I also located and added 9 additional studies published in 2017 and 2018.[15]

For each study I used both the low end of the range of likely values (which I refer to as "minimum estimates") and the high end ("maximum estimates"), as well as the "best" (most likely) estimate. To see how views about climate sensitivity might have changed over time, I divided the studies into two groups based on year of publication: pre-2010 and 2010 onwards. Figure 3.7 shows a histogram with the "best" estimates from these studies.

From the figure, note that the bulk of the studies (115 of the 131) have "best estimates" between 1.5 and 4.5°C, which at least until very recently has been the "most likely" range according to the IPCC. But of course this is still a wide range, and 16 studies have "best estimates" outside this range (as low as 0.5°C and as high as 8°C). We can also get a sense of how views about climate sensitivity changed by comparing the pre-2010 studies with those published 2010 onwards. Both the mean and standard deviation are higher for the more recent studies: 2.77 and 1.03 respectively for the pre-2010 studies, and 2.87 and 1.11 for the later studies.

---

[13] Climate scientists generally agree on the underlying physical laws, but they model (i.e., approximate) those laws in different ways to make their models computationally feasible.

[14] What should we make of different models that give such different probability distributions for climate sensitivity? Should we simply take an average across the models, or weigh the projections of some models more heavily than others? If so, what weights to apply? Others have addressed this problem of "deep" structural uncertainty in climate models, and developed decision rules for making projections from an ensemble of models.

[15] All of the studies that Knutti, Rugenstein, and Hegerl (2017) examined are listed in their paper. The 9 additional studies that I added are: Brown and Caldeira (2017), Krissansen-Totton and Catling (2017), Andrews et al. (2018), Cox, Huntingford, and Williamson (2018), Dessler and Forster (2018), Lewis and Curry (2018), Lohmann and Neubauer (2018), Skeie et al. (2018), and Keery, Holden, and Edwards (2018).

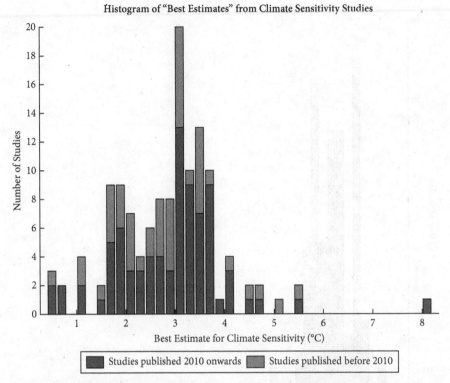

Fig. 3.7 Histogram of Best Estimates of Climate Sensitivity, from 131 studies, of which 47 were published prior to 2010 and 84 from 2010 onwards. The studies are from Knutti, Rugenstein, and Hegerl (2017), supplemented by 9 additional studies published in 2017 and 2018, and listed in Footnote 15 on page 60.

Figures 3.8 shows a histogram for the low end of the range of likely values reported by these studies ("minimum estimates"), and Figure 3.9 shows a histogram for the high end of the range ("maximum estimates"). The bulk of the "minimum estimates" are in the range of 0.5 to 4.0°C, with only three estimates above this range. The bulk of the "maximum estimates" are in the range of 3.0 to 7.0°C, but there are 13 estimates above this range, with seven estimates at 10 to 15°C.

Figures 3.8 and 3.9 tell us that there is a huge amount of uncertainty over climate sensitivity, much more than what is suggested by the five distributions in Figure 3.6. If we ignore the outliers and simply consider the bulk of the "minimum" and "maximum" estimates, we get a range of 0.5 to 7.0°C. Remember that this is a range of "likely" (i.e., probability greater than 66 percent) values, and excludes more extreme values that are unlikely but still possible.

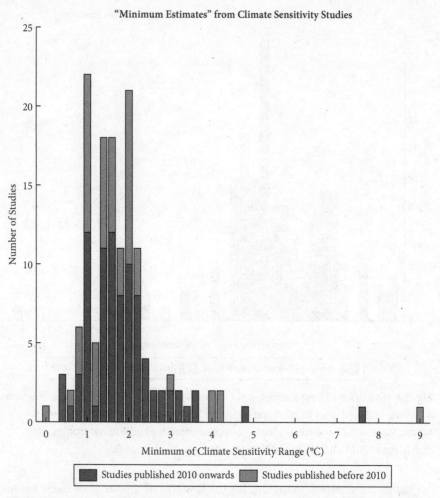

"Minimum Estimates" from Climate Sensitivity Studies

Fig. 3.8 Histogram of Minimum Estimates of Climate Sensitivity from from 143 studies, of which 54 were published prior to 2010 and 89 from 2010 onwards. The studies are from Knutti, Rugenstein, and Hegerl (2017), supplemented by 9 additional studies published in 2017 and 2018, and listed in Footnote 15 on page 60.

Climate scientists have conducted numerous studies that try to estimate climate sensitivity. The individual studies show large ranges of "likely" values, and that range becomes much greater once we account for the dispersion across the different studies. The bottom line: At this point we simply don't know the true value climate sensitivity. And that's unfortunate, because climate sensitivity is a critical determinant of the temperature increases we can expect over the coming decades.

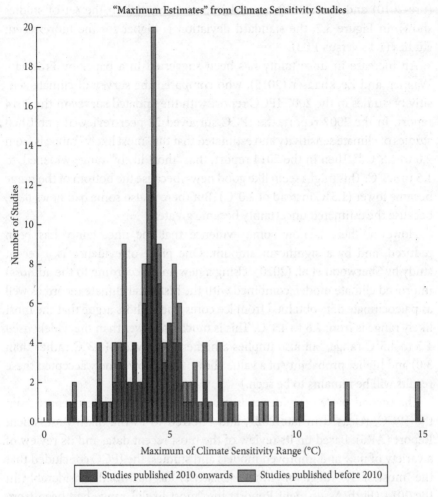

"Maximum Estimates" from Climate Sensitivity Studies

Fig. 3.9 Histogram of Maximum Estimates of Climate Sensitivity from from 143 studies from Knutti, Rugenstein, and Hegerl (2017), supplemented by 9 studies published in 2017 and 2018, and listed in Footnote 15 on page 60.

Has the Uncertainty Been Reduced?

Climate scientists have been busy, publishing hundreds of papers that directly or indirectly relate to climate sensitivity. There is little question that our understanding of the physical mechanisms that underlie climate sensitivity has improved considerably over the past couple of decades. Doesn't this mean that we are now better able to pinpoint the magnitude of climate sensitivity, i.e., that our uncertainty over its true value has been reduced?

Until the most recent (2021) IPCC report was released, we would have had to say that the answer is probably no. This is evident from the earlier (pre-2010) and later (2010 onwards) "best estimates" in the set of studies shown in Figure 3.7; the standard deviation is higher for the more recent studies (1.13 versus 1.03).

An increase in uncertainty has been suggested in a paper by Freeman, Wagner, and Zeckhauser (2015), who compared the survey of climate sensitivity studies in the 2007 IPCC report with the updated survey in the 2014 report. In the 2007 report, the IPCC surveyed 22 peer-reviewed published studies of climate sensitivity and estimated that the "most likely" range is from 2.0 to 4.5°C.[16] Then in the 2014 report, that "most likely" range widened, to 1.5 to 4.5°C. This might seem like good news, because the bottom of the range became lower (1.5°C instead of 2.0°C). But there is also some bad news here, because the estimated uncertainty became greater.

However, there is now some evidence that the uncertainty has been reduced, and by a significant amount. One piece of evidence is a recent study by Sherwood et al. (2020). Using a new and (according to the authors) improved climate model, combined with the historical climate record as well as paleoclimate data obtained from ice cores, the authors argue that the most likely range is from 2.6 to 4.1°C. This is much narrower than the widely used 1.5 to 4.5°C range, but also implies a higher mean value (3.3°C rather than 3.0) and higher probability of a value above 4°C. (How widely accepted these results will be remains to be seen.)

The second piece of evidence is more persuasive: the latest report from the IPCC, released in late 2021, and referred to as the Sixth Assessment Report (AR6). Based on its review of the most recent data, and its review of a variety of new and improved models and studies, the IPCC concluded that the "most likely" range for climate sensitivity has narrowed considerably. In the 2014 (Fifth Assessment Report) the "most likely" range had been from 1.5°C to 4.5°C. In this updated report, the range is from 2.5°C to 4.0°C. (The IPCC also gave as its "best estimate" 3.0°C, which was at the midpoint of the earlier range.)

This reduction in uncertainty over the value of climate sensitivity is very good news. It reflects the fact that the work of climate scientists has given us a better understanding of the physical mechanisms through which increases in the atmospheric CO_2 concentration affect temperature. And now there is reason to hope that ongoing work will give us an even better understanding of the climate system, and a more precise estimate of climate sensitivity.

---

[16] Intergovernmental Panel on Climate Change (2007) also provides a detailed and readable overview of the physical mechanisms involved in climate change, and the state of our knowledge regarding those mechanisms. Each of the individual studies included a probability distribution for climate sensitivity, and by putting the distributions in a standardized form, the IPCC created a graph that showed all of the distributions in a summary form. This is updated in Intergovernmental Panel on Climate Change (2014).

Despite these gains, there is still substantial uncertainty. Keep in mind that a better understanding of the climate system need not by itself mean reduced uncertainty over the value of climate sensitivity. Instead, it can simply provide clarity over *why* there is uncertainty. As an example, the entire field of geology has seen huge gains over the past 50 years. But our greatly improved understanding of the physical mechanisms underlying earthquakes and volcanoes does not mean that we can make more accurate predictions of future earthquakes and volcanic eruptions. A better understanding of how and why earthquakes occur need not translate into more accurate predictions of *when* the next earthquake will arrive. And the field of economics has likewise seen huge gains (please indulge me on that one), so that compared to 50 years ago we now have a better understanding of the workings of individual markets and the economy as a whole. But that better understanding does not mean that we can make more accurate predictions of the timing and severity of the next recession or financial crisis. Instead, we have a better understanding of *why* we can't predict the next recession or financial crisis.

The objective of scientific research is not always to be able to make better predictions. The objective is to better understand what is going on. Research in climate science has given us a much better understanding of what happens when CO_2 is emitted into the atmosphere. To some extent that has helped us better predict what the CO_2 will do to temperature. But it has also helped us understand why their is so much uncertainty in the first place.

Why Is There Uncertainty Over Climate Sensitivity?

The basic problem is that the physical mechanisms that determine climate sensitivity are complex and not fully understood. But most important, the magnitude of climate sensitivity is determined by crucial feedback loops, and we have only rough estimates of the parameter values that determine the strength (and even the sign) of those feedback loops. This is not a shortcoming of climate science; on the contrary, climate scientists have made enormous progress in understanding the physical mechanisms involved in climate change. But part of that progress is a clearer realization that there are limits (at least currently) to our ability to pin down the strength of the key feedback loops.

The problem is easiest to understand in the context of a simple (but widely cited) climate model developed by Roe and Baker (2007). It works as follows. Let S_0 represent climate sensitivity in the absence of any feedback effects. In other words, absent feedback effects, a doubling of the atmospheric CO_2 concentration would cause an increase in radiative forcing that would in turn cause an initial temperature increase of $\Delta T_0 = S_0 °C$. But as Roe and

Baker explain, the initial temperature increase ΔT_0 "induces changes in the underlying processes...which modify the effective forcing, which, in turn, modifies ΔT." As a result, the actual climate sensitivity, S, is given by

$$S = \frac{S_0}{1-f},$$

where f (a number between 0 and 1) is the total feedback factor.[17] So if, for example, $f = 0.95$, then S would equal $S_0/(1 - 0.95) = 20 \times S_0$.

It is important to stress that this is an extremely simplified model of the climate system. A more complete and complex model would incorporate several feedback effects; here they are all being rolled into one. Nonetheless, this simple model allows us to address the key problem: Climate sensitivity is very sensitive to the value of f, but we don't know the value of f. Roe and Baker point out that if we knew the mean and standard deviation of f, denoted by $\bar{f}$ and σ_f respectively, and if σ_f is small, then the standard deviation of S would be proportional to $\sigma_f/(1 - \bar{f})^2$. This tells us that uncertainty over S is greatly magnified by uncertainty over f, and becomes very large if f is close to 1.

For example, suppose our best estimate of f is 0.95, but we believe that could be off by a factor of 0.03, i.e., f could be as low as 0.92 or as high as 0.98. In that case, S could be as low as $(1/.08) \times S_0 = 12.5 \times S_0$ or as high as $(1/.02) \times S_0 = 50 \times S_0$. But $50 \times S_0$ is 4 times as large as $12.5 \times S_0$, so this seemingly small uncertainty over f creates a huge amount of uncertainty over climate sensitivity.

To illustrate the problem further, Roe and Baker assume that f is normally distributed (with mean $\bar{f}$ and standard deviation σ_f), and derive the resulting distribution for S, climate sensitivity. Given their choice of the mean and standard deviation of f, they obtain a distribution for climate sensitivity S, and find that the resulting median and 95th percentile are close to the corresponding numbers that come from averaging across the standardized distributions summarized by the IPCC.[18]

---

[17] In the notation of Roe and Baker (2007), λ_0 is climate sensitivity without feedback effects, and λ is climate sensitivity accounting for feedback effects.

[18] Adding a displacement parameter θ, the Roe-Baker distribution for climate sensitivity is given by:

$$g(S;\bar{f},\sigma_f,\theta) = \frac{1}{\sigma_f\sqrt{2\pi}z^2} \exp\left[-\frac{1}{2}\left(\frac{1-\bar{f}-1/z}{\sigma_f}\right)^2\right],$$

where $z = S + \theta$. Fitting to the distributions summarized by the IPCC, the parameter values are $\bar{f} = 0.797, \sigma_f = .0441$, and $\theta = 2.13$. This distribution is fat-tailed, i.e., declines to zero more slowly than exponentially. Weitzman (2009, 2011, 2014b) has shown that parameter uncertainty can lead to a fat-tailed distribution for climate sensitivity, and that this implies a relatively high probability of a catastrophic outcome, which in turn suggests that the value of abatement is high. Pindyck (2011a) shows that a fat-tailed distribution by itself need not imply a high value of abatement.

This Roe-Baker distribution has become well-known and has been used in several studies to obtain estimates of the SCC, and to analyze how those estimates are affected by uncertainty over climate sensitivity. But it may well understate our uncertainty over climate sensitivity. The reason is that we don't know whether the feedback factor f is in fact normally distributed (and even if it is, we don't know its true mean and standard deviation). Roe and Baker simply assumed a normal distribution to illustrate the implications of uncertainty over the crucial feedback factors. In fact, in an accompanying article in the journal *Science*, Allen and Frame (2007) argued that climate sensitivity is in the realm of the "unknowable," and that considerable uncertainty will remain for decades to come.

3.4.2 The Impact of Climate Change

When assessing climate sensitivity, we at least have the results of a substantial amount of scientific research to rely on, and we can use those results to argue coherently about the probability distributions that they imply. When it comes to predicting the impact of climate change, however, we have much less to go on, and the uncertainty is far greater. In fact, we know very little, if anything, about the impact that higher temperatures and rising sea levels would have on the economy, and on society more generally.

Why is it so difficult to estimate how climate change will affect the economy? One problem is that we have very little data on which to base empirical work. True, we do have data on temperatures in different locations and different periods of time, and we can try to relate changes in temperature to changes in GDP and other measures of economic output. Some research has done just this; we have seen empirical studies that made use of weather data for a large panel of countries over fifty or more years. For example, Dell, Jones, and Olken (2012) demonstrated that the impact of higher temperatures is largely on the growth rate of GDP, as opposed to the level of GDP, and is mostly significant in poor countries.[19] And there have been many more studies that explore how changes in temperature and rainfall affect agricultural output.[20]

But all of these studies suffer from a fundamental problem: They relate changes in *weather* to changes in GDP or agricultural output, and *weather is not the same as climate*. The weather in any location—temperature, rainfall,

---

[19] See Dell, Jones, and Olken (2014) for an overview of this line of research.

[20] One of the earliest such studies is Mendelsohn, Nordhaus, and Shaw (1994). A more recent study is Deschênes and Greenstone (2007). For overviews, see Auffhammer et al. (2013) and Blanc and Schlenker (2017).

humidity, etc.—changes from week to week and month to month. But the climate—which describes the average temperature and rainfall that we can expect in any particular week or month of the year—changes very slowly (if at all). An unexpectedly hot summer might indeed reduce the size of that year's wheat or corn harvest, but the impact of a gradual change in climate (in which average expected temperatures rise) might have a very different (and probably lower) impact, because farmers will shift what and where they plant. Finally, the observed changes in temperature that are used in these studies are relatively small—not the 4°C or more of warming that many people worry about.

A second problem is that there is little or nothing in the way of economic theory to help us understand the potential impact of higher temperatures. We have some sense of how higher temperatures might affect agriculture, and indeed, most of the empirical work that has been done is focused on agriculture. But we also know that losses of agricultural output in some regions of the world (e.g., near the equator) might be offset by increased output in other regions (e.g., northern Canada and Russia). Furthermore, agriculture is a small fraction of total economic output: 1 to 2 percent of GDP for industrialized countries, 3 to 20 percent of GDP for developing countries. Beyond agriculture, it is difficult to explain, even at a heuristic level, how higher temperatures will affect economic activity.

A third problem is that climate change will occur slowly, which means there is considerable potential for adaptation on the part of the people and firms that might be impacted. In the case of agriculture, we already saw substantial adaptation on the part of farmers in the United States during the 19th century as settlers moved west and had to adapt crops to new and very different climatic conditions. (This history of adaptation in agriculture is discussed in detail in Chapter 7.) Flooding is a potential hazard of climate change if sea levels rise considerably, but here, too, we have seen adaptation in the past (with the dikes of Holland perhaps the best known example). This does not mean that adaptation will eliminate the impact of climate change—it is simply another complicating factor that makes it very difficult to estimate the extent of the losses we should expect.

It may be that the relationship between temperature and the economy is not just something we don't know, but something that we *cannot* know, at least for the time horizon relevant to the design and evaluation of climate policy. As discussed earlier, some researchers have come to the conclusion that climate sensitivity is in this category of the "unknowable." Yet, for the reasons just discussed, the impact of climate change is even less "knowable" than climate sensitivity.

Models and Damage Functions

The impact of climate change is a key element in the integrated assessment models (IAMs), that have proliferated over the past couple of decades. These models "integrate" a description of GHG emissions and their impact on temperature (a climate science model) with a description of how changes in climate affect output, consumption, and other economic variables (an economic model). Although they have been widely used to estimate the social cost of carbon (SCC), they have serious flaws, which I have discussed in detail elsewhere (Pindyck (2013a, 2017b)). Here I focus on the economic part of these models, i.e., how they describe the impact of climate change.

Most IAMs relate the temperature increase ΔT to GDP through a damage function or "loss function" $L(\Delta T)$, with $L(0) = 1$ and as ΔT gets bigger, $L(\Delta T)$ gets smaller (i.e., $dL(\Delta T)/d\Delta T < 0$). The idea here is that GDP at time t is $GDP_t = L(\Delta T_t)GDP_t'$, where GDP_t' is what GDP would be if there were no increase in temperature, and $1 - L(\Delta T_t)$ is the reduction in GDP_t' due to higher temperatures. For example, if ΔT is 3°C and $L(3) = .95$, that would mean that the 3°C temperature increase will reduce GDP by $1 - .95 = .05$, i.e., by 5 percent.

Different IAMs have different loss functions. The widely used Nordhaus (2008) DICE (Dynamic Integrated Climate and Economy) model has the following inverse-quadratic loss function:

$$L(\Delta T) = 1/[1 + \pi_1 \Delta T + \pi_2 (\Delta T)^2]. \tag{3.1}$$

On the other hand, Weitzman (2009) suggested the exponential-quadratic loss function:

$$L(T) = \exp[-\beta(\Delta T)^2], \tag{3.2}$$

which allows for greater losses when the temperature increase ΔT is large.

Which of these two loss functions is better, or more accurate? We can't say. The problem is that neither of the loss functions is based on any economic (or other) theory. Nor are the loss functions that appear in other IAMs.[21] They are just arbitrary functions, made up to describe how GDP goes down when temperature goes up.

---

[21] In addition to the DICE model, the U.S. Government's Interagency Working Group ran simulations of two other IAMs, with a range of parameter values and discount rates, to estimate the SCC. The other two IAMS were PAGE (Policy Analysis of the Greenhouse Effect) and FUND (Climate Framework for Uncertainty, Distribution, and Negotiation), which are described in Hope (2006) and Tol (2002a,b). For an explanation of how the models were used to estimate the SCC, see Greenstone, Kopits, and Wolverton (2013) and Interagency Working Group on Social Cost of Carbon (2013).

Should we conclude that IAM developers were negligent and ignored economic theory when building their models? Not at all. There is no economic (or other) theory that can tell us what the loss function $L(\Delta T)$ should look like.[22]

Furthermore, suppose some economic theory did tell us that the inverse-quadratic loss function used in the DICE model (and shown above) is indeed a credible description of the impact of higher temperatures. Then the question would be how to determine the values of the parameters π_1 and π_2. Theory can't help us, nor is data available that could be used to estimate or even roughly calibrate the parameters. As a result, the choice of values for these parameters is essentially guesswork. One approach is to select values such that $L(\Delta T)$ for ΔT in the range of 2°C to 3°C is consistent with common wisdom regarding the damages that are likely to occur for small to moderate increases in temperature. Most modelers choose parameters so that $L(1)$ is close to 1 (i.e., no loss), $L(2)$ is around .99 or .98, and $L(3)$ is around .98 to .95. But of course we don't know whether this common wisdom is correct; we don't know whether a 3°C increase in temperature would cause a 2 to 5 percent loss of GDP, and we probably won't know until we actually experience such an increase in temperature.

The bottom line is that the damage functions used in most of the models lack much in the way of a theoretical or empirical foundation. That might not matter much if we are looking at a temperature increase of only 2°C, because there is a rough consensus (perhaps completely wrong) that damages will be small to moderate at that level of warming. The problem is that these damage functions tell us nothing about what to expect if temperature increases are larger, e.g., 4°C or more.[23]

I do not want to give the impression that economists know nothing about the impact of climate change. On the contrary, considerable work has been done on specific aspects of that impact, especially with respect to agriculture.

---

[22] If anything, we would expect higher temperatures to affect the *growth rate* of GDP, and not the level. Why? First, some effects of warming will be permanent; e.g., destruction of ecosystems and deaths from weather extremes. A growth rate effect allows warming to have a permanent impact. Second, the resources needed to counter the impact of warming will reduce those available for R&D and capital investment, reducing growth. Third, there is some empirical support for a growth rate effect. Using data on temperatures and precipitation over 50 years for a panel of 136 countries, Dell, Jones, and Olken (2012) have shown that higher temperatures reduce GDP growth rates but not levels. See Pindyck (2011*b*, 2012) for further discussion and an analysis of the policy implications of a growth rate versus level effects. Note that a climate-induced catastrophe, on the other hand, would reduce both the growth rate and level of GDP.

[23] Some modelers are aware of this problem. Nordhaus (2008) states (p. 51): "The damage functions continue to be a major source of modeling uncertainty in the DICE model." To get a sense of the wide range of damage numbers that come from different models, even for temperature increases of 2 or 3°C, see Tol (2018). Stern (2013) argues that IAM damage functions ignore a variety of potential climate impacts, including possibly catastrophic ones, and Diaz and Moore (2017) critique IAM damage functions in terms of parameter uncertainty. Burke et al. (2015) explores uncertainty over climate impacts, but only arising from the uncertainty over climate change itself.

Some of the studies of agricultural impacts include Deschênes and Greenstone (2007) and Schlenker and Roberts (2009). A study that focuses on the impact of climate change on mortality, and our ability to adapt, is Deschênes and Greenstone (2011). And recent studies that use or discuss the use of detailed weather data include Fisher et al. (2012) and Auffhammer et al. (2013). These are just a few examples; the literature is large and growing.

Statistical studies of this sort will surely improve our knowledge of how climate change might affect the economy, or at least some sectors of the economy. But the data used in these studies are limited to relatively short time periods and small fluctuations in temperature and other weather variables—the data do not, for example, describe what has happened over 20 or 50 years following a 4°C increase in mean temperature. Given this limitation, these studies do not enable us to specify and calibrate damage functions of the sort used in IAMs. (In fact, those damage functions have little or nothing to do with the detailed econometric studies related to agricultural and other specific impacts.) In part because of the very limited data that are available, the estimates of the impacts exhibit huge variation. This can be seen from two recent surveys: Tol (2018) and Nordhaus and Moffat (2017). Both of these surveys show wide variation in various estimates of impacts. And even then, the surveys apply only to impacts from relatively small temperature changes, mostly less than 3°C.

3.4.3 A Catastrophic Outcome

It may well turn out that over the coming decades climate change and its impact will be mild to moderate. Given all of the uncertainties over climate sensitivity and climate impacts, this might turn out to be the case even if little is done to reduce GHG emissions. And if we were *certain* that this will be case, it would imply that the social cost of carbon is quite low, and we can relax and stop worrying about climate change.

But we are *not* certain that the outcome will be so favorable. There is a possibility of an extremely unfavorable outcome, one that we could call *catastrophic*. The kind of outcome I am referring to is not simply a very large increase in temperature, but rather a very large economic effect, in terms of a decline in human welfare, from whatever climate change occurs. The IAMs and related models that have been used to estimate the Social Cost of Carbon have little or nothing to tell us about such outcomes. This is not surprising; as I explained, the damage functions in the models, which anyway are ad hoc, are calibrated to give small damages for small temperature increases, and can say nothing meaningful about the kinds of damages we should expect for temperature increases of 4°C or more. And that's unfortunate, because it is the possibility of a catastrophic outcome that really drives the SCC and matters for climate policy.

What do we mean by a "catastrophic outcome?" For climate scientists, it usually means a high temperature outcome. How high? There is no fixed rule here. Almost everyone working on climate change would agree that a 5°C or 6°C temperature increase by 2100 would be in the realm of the catastrophic. A temperature increase this large might result if the climate system reaches a *tipping point* as the CO_2 concentration keeps increasing. By "tipping point," I am referring to a runaway feedback phenomenon in which (for example) warming causes the release of more GHGs (perhaps from the thawing of the permafrost), which causes more warming, which causes the release of still more GHGs, and so on.

Putting aside the difficulty of estimating the probability of extreme warming, what matters in the end is not the temperature increase itself, but rather its impact. Would that impact be "catastrophic," and might a smaller (and more likely) temperature increase, perhaps 3°C, be sufficient to have a catastrophic impact? Again, opinions vary. Some have argued that even a 2°C temperature increase would be catastrophic. For example, CarbonBrief, an interactive collection of 70 peer-reviewed climate studies that show how different temperatures are projected to affect the world, has suggested that 2°C of warming could result in a permanent reduction in global GDP of 13 percent.[24]

Why does the possibility of a catastrophic outcome matter so much for climate policy? Because even if such an outcome has a low probability of occurring, a severe loss of GDP (broadly interpreted) can push up the SCC considerably, justifying a large carbon tax (or equivalent emission abatement policy). A mild to moderate outcome, on the other hand, is something to which society can respond, in part through adaptation, at a relatively low cost. This means that to a large extent, climate policy has to be based on the (small) likelihood of an extreme outcome.

So how likely is a catastrophic outcome, and how catastrophic might it turn out to be? How high can the atmospheric CO_2 concentration be before the climate system reaches a "tipping point," and temperatures rise rapidly? I wish this weren't the case, but the fact is we simply don't know. We don't know where a tipping point, if there even is one, might lie, and what the impact of a large temperature increase might be. Furthermore, it is difficult to see how

---

[24] The CarbonBrief website is https://www.carbonbrief.org/.

answers to these questions will become clear in the next few years, despite all of the ongoing research on climate change. The likelihood and impact of a catastrophic outcome may simply be in the realm of the "unknowable." But that doesn't mean we should ignore the possibility. On the contrary, the chance of a climate catastrophe should be front and center when we think about climate policy, as I explain in the next chapter.

3.5 Further Readings

In this chapter I provided a brief overview of what we know and don't know about climate change, including the nature and extent of the uncertainty. The literature is vast, but for those readers who would like to learn more, I would suggest (as a start) the following books and articles, some of which I recommended at the end of Chapter 1:

- Two books that provide nice introductions to climate change, with a focus on the science, are *Climate Change: What Everyone Needs to Know* by Romm (2018), and *Global Warming: The Complete Briefing*, by Houghton (2015). For another introduction to climate change, but with a focus on the economics, see Heal (2017a). The connection between GDP growth and climate change is discussed by Stock (2019), who describes statistical approaches to estimating the relationship.
- For a thorough and detailed treatment of what we know and don't know about climate change, its impacts, and possible mitigation strategies, see the three-volume report of the IPCC, and the IPCC's 2018 Special Report on the possible impact of a temperature increase above 1.5°C: Intergovernmental Panel on Climate Change (2014, 2018). Much of what is in this chapter also appears in Pindyck (2021).
- Kopits, Marten, and Wolverton (2013) provides a nice overview of why the possibility of a catastrophic climate outcome can be the main driver of the social cost of carbon, and the implications for policy analysis. For discussions of climate uncertainty, and especially how uncertainty over the likelihood of a catastrophic outcome complicates the analysis of climate policy, see Pindyck (2013b) and Heal and Millner (2014). Hawkins and Sutton (2009) show how climate change uncertainty can be better understood by breaking it down according to the source.
- A growing number of studies relate changes in temperature to changes in GDP and other measures of economic output, including agricultural output. Dell, Jones, and Olken (2014), Auffhammer et al. (2013), and Blanc

and Schlenker (2017) provide overviews of some of this research. There is a very wide range of estimates of the impact of rising temperatures (and climate change more generally); for two recent recent surveys, see Tol (2018) and Nordhaus and Moffat (2017). Also, Auffhammer (2018) and Kolstad and Moore (2020) provide nice overviews of why it is so difficult to quantify the economic damages likely to result from climate change, some of the statistical methods that have been used, and some of progress that has been made.

- I explained that the Social Cost of Carbon (SCC) is a useful metric for the design of a carbon tax or other policy to reduce CO_2 emissions. Why and how should we tax externalities such as CO_2 emissions? For an introduction to the basics of environmental policy, see Chapter 18 of Pindyck and Rubinfeld (2018). Two excellent textbooks that provide a thorough treatment of environmental economics and policy are Kolstad (2010) and Phaneuf and Requate (2017).

- For recent estimates of the SCC, see National Academy of Sciences (2017). Some have argued that the use of the SCC is problematic because of the non-marginal nature of damages; see, e.g., Morgan et al. (2017). I recently conducted a survey of several hundred experts in climate science and climate economics to get their opinions on the SCC. The results showed considerable heterogeneity across experts, and wide variation in the implied SCC numbers. See Pindyck (2019) for details.

- The SCC will depend critically on the discount rate used to put future costs and benefits in present value terms. Given that most of the benefits from GHG emission reductions will occur in the distant future, but costs accrue now, a high discount rate will make the SCC relatively low. For an introduction to discounting, see Gollier (2001, 2013) and Frederick (2006).

4
The Role of Uncertainty in Climate Policy

As I explained in the Introduction, many of the books, articles, and press reports that we read make it seem that we know a lot more about climate change and its impact than is actually the case. Likewise, commentators and politicians often make statements of the sort that if we don't take immediate action and sharply reduce CO_2 emissions, the following things will happen, as though we actually knew what will happen. Rarely do we read or hear that those things *might happen*; instead we're told they *will happen*.

This shouldn't come as a surprise. We humans prefer certainty to uncertainty, and feel uncomfortable when we don't know what lies ahead. In addition, most of us have trouble even understanding concepts involving probabilities.[1] Most people prefer to hear or read statements of the sort "By 2050 the temperature will rise by X°C, sea levels will rise by Y meters, and as a result GDP will fall by Z percent," as opposed to "there is a 10-percent chance that the temperature will rise by X°C." Many people ignore the fact, or find it hard to accept, that even if we could accurately predict future GHG emissions, we don't know—and at this point can't know—by how much the temperature or sea levels will rise. And even if we could accurately predict how much the temperature and sea levels will rise, we don't know what the impact would be on GDP or other measures of economic and social welfare. As discussed in the previous chapter, the simple fact is that the "climate outcome," by which I mean the extent of climate change and its impact on the economy and society more generally, is far more uncertain than most people think.

In this chapter I turn to the policy implications of this uncertainty. You might think so much uncertainty should lead us to wait and see what happens, rather than try to sharply reduce emissions right away. After all, if we don't know how much the climate will change, and we don't know what the impact

---

[1] As a result, economists are often pushed to make point predictions, even though they know (but may not want to admit) that they have little or no basis for the prediction. See Manski (2020) for a nice discussion of this problem.

Climate Future: Averting and Adapting to Climate Change. Robert S. Pindyck, Oxford University Press.
© Oxford University Press 2022. DOI: 10.1093/oso/9780197647349.003.0004

of climate change will be, why take costly actions now? That is indeed the argument made by many of the people who oppose the imposition of carbon taxes or other measures to reduce emissions. But that argument is wrong, and actually gets things backwards. As we will see, *the uncertainty itself can lead us to act now*. Why? Because with uncertainty, and especially with the possibility of an extreme outcome, we need insurance.

At this point we don't even know the kinds of climate policies that countries will adopt, and thus the extent to which the world will reduce CO_2 emissions. Of course, whatever happens to CO_2 emissions, it may turn out that we will be lucky and any climate change and its impact will be small. But we might not be so lucky; we might experience a catastrophic climate outcome, with costs to society that are huge. This uncertainty doesn't mean we should ignore the problem and take no action. Instead, we should take action now as insurance against the possibility of very high costs in the future.

Think about this in the context of homeowner buying insurance. You don't know whether your home will be damaged by a fire, flood, or a falling tree during the coming years, never mind how much damage might result from such an event. But that doesn't mean you shouldn't buy insurance for your home, and simply wait to see what happens. On the contrary, a prudent homeowner will buy enough insurance to cover the potential cost of an adverse event, even if there is only a small chance of that event occurring.

Uncertainty over the climate outcome has other implications as well. Consider the *irreversibilities* that are an inherent part of climate policy (and environmental policy more generally). It has been long understood that environmental damage can be irreversible, which can lead to a more "conservationist" policy than would be optimal otherwise. Thanks to Joni Mitchell, even non-economists know that if we "pave paradise and put up a parking lot," paradise may be gone forever. And if the value of paradise to future generations is uncertain, the benefit from protecting it today should include an "option value," which pushes the cost-benefit calculation towards protection.

But there is a second kind of irreversibility that works in the opposite direction: Protecting paradise imposes *sunk costs* on society. By "sunk costs," we mean costs that cannot be recovered, so that the expenditure is irreversible. If paradise includes clean air and clean water, protecting it could imply sunk cost investments in abatement equipment, and an ongoing flow of sunk costs for more expensive production processes. In other words, protecting paradise requires irreversible expenditures, money that cannot be recovered in the future. This kind of irreversibility would lead to policies that are *less* "conservationist" than they would be otherwise, i.e., they would push the cost-benefit calculation away from protection.

Which of these two irreversibilities applies to climate policy? Both. Given that they work in opposite directions, which one is more important? It's hard to say. Read on and you'll understand how these irreversibilities work, and although both are important, why we can't say which one is *more* important.

4.1 Implications of Uncertainty

I have argued that when it comes to future CO_2 emissions and their accumulation in the atmosphere, we have a good understanding of what is going on. Yes, there is uncertainty—over long-run changes in GDP, changes in carbon intensity, and CO_2 dissipation rates—but the uncertainty is limited and can be bounded. When it comes to climate sensitivity—and thus long-run changes in temperature—the uncertainty is far greater, so that mild versus severe temperature outcomes can differ by a factor of three or more. And finally, when it comes to the impact of higher temperatures, we have very little in the way of theory or data to go on, so our projections boil down to guesswork. That in turn means that the range of possibilities is huge.

These uncertainties make the design and analysis of climate policy very different from most other problems in environmental economics. Most environmental problems are amenable to standard cost-benefit analysis. Determining the limits to be placed on sulfur dioxide and nitrous oxide emissions from coal-burning power plants is a good example. These emissions can be very harmful to the health of people living downwind, and also cause acidification of lakes and rivers, harming fish and other wildlife. We would like to limit these emissions, but doing so is costly because it would raise the price of the electricity produced by the power plant. Sulfur dioxide emissions are typically reduced by installing expensive "scrubbers" that remove the pollutant from the exhaust of the plant, or alternatively by burning (more expensive) low-sulfur coal.[2] On the other hand, the benefit of reducing emissions is a reduction in the health problems that they cause, and less damage to lakes and rivers.

So how should we decide the extent to which power plant emissions should be reduced? We compare the cost of any particular emission reduction to the resulting benefit, and consider reducing emissions further if the cost is less than the benefit. There will be uncertainties over the costs and benefits of any candidate policy, but the characteristics and extent of those uncertainties will

---

[2] The use of coal to generate electricity has been dropping dramatically, largely because of the availability of cheap natural gas, and more recently because of the growth of solar and wind power.

usually be well understood, and comparable in nature to the uncertainties involved in many other public and private policy or investment decisions. Of course, economists can (and will) argue about the details of the analysis. But at a basic level, we're in well-charted territory and we think we know what we're doing. If we come to the conclusion that a policy to reduce sulfur dioxide emissions by some amount is warranted, that conclusion will be seen—at least by most economists—as defensible and reasonable.

But as I have explained, this is not the case when it comes to climate change. Climate policy is controversial, in part because the uncertainties complicate the argument for more or less stringent emissions abatement. There is disagreement among both climate scientists and economists over the likelihood of alternative climate outcomes, especially catastrophic outcomes. There is also disagreement over the framework that should be used to evaluate the potential benefits from an abatement policy. The discount rate to be used to compare future benefits with present costs is particularly important in this regard, because most of the benefits will occur in the far future. These disagreements make climate policy much less amenable to standard cost-benefit analysis.

The bottom line is that climate policy is complicated by the huge amount of uncertainty that we face over the extent and impact of future climate change. Furthermore, although the uncertainty applies to the whole range of possible temperature increases and possible impacts, it is especially problematic when it comes to catastrophic outcomes. So what should we do? Is there a way to properly account for this uncertainty in our models of climate change? How should we handle the possibility of a catastrophic outcome? And how can we account for the insurance value of early action, and the conflicting irreversibilities inherent in climate policy?

4.1.1 The Treatment of Uncertainty

For the past couple of decades, the development of IAMs and related models has been a growth industry. Dozens of large models (and dozens more of smaller ones) have been built and used to make forecasts of changes in temperature, sea levels, and other aspects of climate, as well as the economic impacts of those changes. But the fact that there is so much uncertainty over climate change and its impact makes the value of these forecasts (and perhaps the models themselves) questionable, to say the least. What can be done? Should we simply declare that we don't know what is going to happen, and proceed accordingly? If so, what does "proceed accordingly" involve?

Most of the people that build and use models do not think we should abandon them, and argue that there are ways to account for uncertainty, so that the models can still be useful. One approach is to incorporate uncertainty in the parameters of the model. In the previous chapter, for example, we introduced the loss function of eqn. (3.1), which has two parameters (π_1 and π_2). So, to account for uncertainty over the impact of higher temperatures, we might treat these two parameters as random variables, and see how that affects the range of likely impacts. That's the basis of Monte Carlo simulation, which is discussed below.

A second (but less widely used) approach is to build models in which the parameters might or might not be assumed to be known, but uncertainty is built into the workings of the model. For example, investment in "green technologies" will depend on how risky are the returns to that investment, which depends on the overall risk in the economy as well as technology-specific risk. And a third approach is take a model with uncertain parameters, and summarize the uncertainty in the form of "best-case" and "worst-case" outcomes.

Parameter Uncertainty: Monte Carlo Simulation

The developers of IAMs and related models acknowledge that there is uncertainty over parameters and even the functional forms of some of their equations. What, then, can be done about it? One approach that modelers have used is *Monte Carlo simulation*. Rather than treat each parameter in the model as a fixed and known number, a probability distribution is assigned to the parameter. Where does the probability distribution come from? It is chosen by the modeler, and represents the modeler's view about the nature of the uncertainty over that parameter. (The standard deviation of the probability distribution, again chosen by the modeler, would reflect the perceived extent of the uncertainty.)

A model might have 10 or 20 parameters that are viewed as uncertain, so there might be 10 or 20 probability distributions (one for each uncertain parameter). Then the model would be run over and over again, perhaps 100,000 or more times, and for each run the probability distributions are used to obtain random draws for each parameter. From the 100,000 runs we obtain a distribution (including a mean and standard deviation) for the output variable of interest, such as temperature or lost GDP at the end of the century.[3]

---

[3] Nordhaus (2018) provides a clear example of the application of Monte Carlo simulation to an IAM. MIT's Joint Program on the Science and Policy of Global Change stressed the importance and implications of uncertainty early on in their modeling work, and acknowledged that there is no basis for including a damage function in the models.

This makes sense, and indeed, Monte Carlo simulation has been used extensively in many scientific applications. But in the context of climate change, does it really tell us much about the nature of the uncertainty and its implications for policy? Unfortunately, the answer is no. The problem is deciding on the probability distributions that would be applied to each of the parameters. For most of the parameters, we simply don't know the correct probability distributions (just as we don't know the parameter values), and different distributions can yield very different results for expected outcomes.[4]

To make matters worse, we don't even know the correct functional forms for some of the key relationships. This is particularly a problem when it comes to the impact of climate change. As I mentioned earlier, the loss function used in the Nordhaus DICE model is a simple inverse quadratic:

$$L(\Delta T) = 1/[1 + \pi_1 \Delta T + \pi_2 (\Delta T)^2],$$

where ΔT is the anthropomorphic increase in temperature and $L(\Delta T)$ is the reduction (i.e., the loss) in GDP and consumption for any value of ΔT. But remember that this loss function is completely hypothetical; it is not derived from theory or data. Furthermore, even if this inverse quadratic function were somehow the true loss function, there is no theory or data that can tell us the correct values for the parameters π_1 and π_2, the correct probability distributions for those parameters, or even the correct means and variances.

To illustrate, suppose we (somehow) chose probability distributions for π_1 and π_2. A Monte Carlo simulation would then give us the expected loss $L(\Delta T)$ for any particular temperature increase ΔT. But suppose we then come to believe that damages are likely to rise very rapidly as the temperature increases, more rapidly than the inverse quadratic would indicate. This might lead us to conclude that the loss function should be different, perhaps an inverse cubic rather than quadratic. For example, we might decide that the following loss function is preferred:

$$L(\Delta T) = 1/[1 + \pi_1 \Delta T + \pi_2 (\Delta T)^3].$$

The Monte Carlo simulation will now give us a very different (and larger) expected loss. Likewise, one might argue that we are using the wrong probability distributions for π_1 and π_2, or that we have the correct distributions

---

[4] In Pindyck (2013b), I took three different but plausible distributions for temperature change: a Gamma distribution, a Frechet distribution (also called a Generalized Extreme Value, Type 2 distribution), and the distribution derived by Roe and Baker (2007). I calibrated all three distributions so they have the same mean and variance, and I demonstrated that they imply very different values for the social willingness to pay (WTP) to avoid an increase in temperature.

but the wrong means and/or variances for the distributions. Changing the probability distributions or the means and variances will also result in a very different estimate of the expected loss.

Again, Monte Carlo simulation can be a powerful tool for incorporating uncertainty in a model, and is widely used. But it is useful only when applied to a model that has a strong theoretical and empirical foundation, and has parameters for which the probability distributions are well understood and empirically supportable. In the case of climate change, however, we know as little about the correct probability distributions as we do about the damage function to which they are being applied. What can we expect to learn from assigning arbitrary probability distributions to the parameters of an arbitrary function and running Monte Carlo simulations? Unfortunately, not much. The basic problem is simple: If we don't understand how A affects B, but we create some kind of model of how A affects B, running Monte Carlo simulations of the model won't make up for our lack of understanding.[5]

Other Approaches to Incorporating Uncertainty

Another approach is to build uncertainty into the workings of the model. Examples of this are the models in Cai and Lontzek (2019), Rudik (2020), and van den Bremer and van der Ploeg (2021). In those models a variety of important parameters are treated as uncertain, but rather than conduct a Monte Carlo simulation (in which probability distributions for the parameters must be specified), the model is solved for a range of different parameter values. The distinguishing feature, however, is that the dynamic evolution of certain variables is explicitly random. In the Cai and Lontzek (2019) model, for example, economic growth, which is a driver of future carbon emissions, is treated as a random process. This is realistic because future economic growth is indeed inherently uncertain. And with economic growth uncertain, the SCC that the model generates becomes partly random, which means that the future SCC is uncertain.

This is a step forward, because it makes the uncertainty itself an explicit part of the model. In the case of economic growth, for example, it means describing the nature and extent of the uncertainty, and thus how it affects future CO_2 emissions. But while this is a step forward, it still doesn't deal with the fact that we don't know the correct functional forms for some of the key relationships in the model.

---

[5] As Mervyn King, the former Governor of the Bank of England put it (in a very different context): "...if we don't know what the future might hold, we don't know, and there is no point pretending otherwise" (King (2016)). The models (and the users of the models) sometimes pretend otherwise.

Still another approach, taken by Hassler, Krusell, and Olovsson (2018) among others, is to use a model with uncertain parameters to estimate "best-case" and "worst-case" outcomes. The "best-case" outcome is generated using the most favorable (but plausible) parameter values, and the "worst-case" outcome uses the most unfavorable (but plausible) parameter values. An advantage of this approach is that it can shed light on the role of irreversibilities. We would be sorry if we spend a great deal of money now to reduce CO_2 emissions and then, in another 20 or 30 years, learn that climate change is much less of a problem than we thought. And by the same token, we would be sorry if we do very little now to reduce CO_2 emissions, which continue to build up in the atmosphere, and then, in another 20 or 30 years, learn that we are about to face a "worst-case" climate outcome, i.e., one that is catastrophic.

But there is a problem here: Is the "best-case" outcome indeed the most favorable one that we can reasonably expect? And is the "worst-case" outcome indeed the least favorable one? These questions are hard to answer because the "best-case" and "worst-case" outcomes are based on a model (or perhaps several models) that have limited theoretical and empirical support. On the other hand, this approach at least provides some estimates of the range of possible outcomes, and thus the extent of the uncertainty we face. And that is useful because it can help us evaluate the role and importance of irreversibilities and insurance value in the design of climate policy.

4.1.2 How Does Uncertainty Affect Climate Policy?

By this point I hope I've convinced you that when it comes to climate change, we live in a world of extreme uncertainty. In fact, there is uncertainty—or at least considerable disagreement—over the nature and extent of the uncertainty itself. Given this uncertainty, what should we do? Feeling frustrated and running away from the problem is not the best option. Nor is it a good idea to immediately and completely stop producing fossil fuels of any kind, scrap our cars, and turn off the lights. A policy in the middle might be better, but just what should that policy be? And how should the uncertainties we face drive that policy?

At the beginning of this chapter, I explained that uncertainty can affect policy in two ways. First, it creates a value to insurance (against a very bad outcome). That value of insurance can lead us to act sooner and adopt a more stringent emission abatement policy than we would otherwise. Second, there are irreversibilities that are inherent in climate policy, and these irreversibilities can affect policy through their interaction with uncertainty. However, the net effect of these irreversibilities is unclear. As we will soon see, we can't say for sure whether they should lead to a more aggressive or less aggressive climate policy.

4.1.3 The Value of Climate Insurance

Uncertainty over climate change creates insurance value in two different ways, and it is important to keep them clear:

(1) First, it occurs through the "damage function," i.e., the loss of GDP resulting from any particular temperature increase. Although the impact of any increase in temperature is highly uncertain, it is very likely that the damage function becomes increasingly steep as the temperature change becomes larger. Put another way, going from 3°C of warming to 4°C is likely to cause a much larger reduction in GDP than going from 1°C to 2°C. As the temperature increase becomes larger, damages become more severe and adaptation becomes more difficult, so the damage from an additional 1°C of warming becomes larger.

(2) The second way that uncertainty creates insurance value is through social risk aversion. Risk aversion refers to a preference for a sure outcome rather than a risky outcome, even if that risky outcome has the same (or even greater) expected value as the sure outcome. We do not know what the "correct" social welfare function is, but we expect it to exhibit at least some risk aversion. Why? Because most of the people that make up society tend to be risk-averse. This means that society as a whole would pay to avoid the risk of a very bad climate outcome.

The Damage (or Loss) Function

To understand how uncertainty, combined with an increasingly steep damage function, creates a value of insurance, we'll use a very simple example. We will consider a single point in the future, say the year 2050, and we will ignore the issue of discounting future costs and benefits. For purposes of this illustrative example, I will assume that the percentage loss of GDP resulting from a temperature increase ΔT is given by the following equation:

$$L(\Delta T) = 1 - \frac{1}{1 + .01(\Delta T)^2}. \tag{4.1}$$

Eqn. (4.1) says that $L(0) = 0$, i.e., with *no* temperature increase, there would be no loss of GDP. It also says that $L(2) = 0.04$, i.e., a 2°C temperature increase would result in a loss of 4 percent of GDP, $L(4) = 0.14$, i.e., a 4°C temperature increase would result in a loss of 14 percent of GDP, $L(6) = 0.26$, i.e., a 6°C temperature increase would result in a 26 percent loss of GDP, and so on. Note that each additional 2° increase in temperature results in a larger and larger additional loss, which is what we mean by "an increasingly steep damage function."

First, suppose we know for certain that in 2050 the global mean temperature will have increased by 2°C. And suppose we know for certain that this 2°C temperature increase will cause a 4 percent drop in GDP, compared to what GDP would be without the higher temperature, just as equation (4.1) says. What percentage of GDP should we be willing to sacrifice to avoid this temperature increase? Up to 4 percent. Hopefully, we could avoid the temperature increase at a cost that is *less* than 4 percent of GDP (perhaps by developing and making use of new energy-saving technologies). But if we had to, we'd be willing to sacrifice up to a maximum of 4 percent of GDP.

Now, suppose there is uncertainty over the temperature increase. We think that the temperature might not increase at all, or that it might increase by 4°C, with each outcome having a 50 percent probability. What is the *expected value* of the temperature increase? It is $(0.5)(0) + (0.5)(4) = 2°C$. So the expected (or average) size of the temperature increase is the same 2°C as it was in the first case, but now there is uncertainty—it might be zero and it might be 4°C. Does this uncertainty change things?

How bad would a 4°C temperature increase be in terms of its impact on GDP? Would it reduce GDP by 8 percent, i.e., twice the 4 percent drop we said would occur with a 2°C temperature increase? No, we would expect the impact on GDP to be much larger than 8 percent; the damage caused by higher temperatures is likely to rise more than proportionally. Why? Because 4°C of warming is much more likely to cause substantial increases in sea levels (for example by melting the Antarctic ice sheets), substantial damage to crops, and substantial increases in the transmission of contagious diseases. We don't know what the actual impact will be, but we expect it will be more than twice as bad as the impact of a 2°C temperature increase. Using equation (4.1), we will assume it causes a 14 percent drop in GDP. In this case, what percentage of GDP should we be willing to sacrifice to avoid the *possibility* of a 4°C temperature increase?

To answer this, let's consider the *expected size* of the impact on GDP. The expected size of the temperature increase is still 2°C, and we said the impact of a 2°C temperature increase would be 4 percent of GDP. But the expected impact of a fifty-fifty chance of no temperature increase and a 4° temperature increase is *greater* than 4 percent of GDP. It is $(0.5)(0) + (0.5)(14) = 7$ percent of GDP. That says that we should be willing to sacrifice up to 7 percent of

GDP to avoid the 50 percent chance of a 4° temperature increase. (Once again, hopefully we can avoid the temperature increase at a cost that is *less* than 7 percent of GDP, but if we had to, we'd be willing to sacrifice up to that amount.) Given that the expected temperature increase is still 2°C, why would we be willing to sacrifice so much more? Because the 4°C increase in temperature, which admittedly has only a 50 percent chance of occurring, would be so much more damaging.

Let's take this one more step. Suppose there is a 75 percent probability that there will be no temperature increase, and just a 25 percent chance of an 8° temperature increase. And suppose that an 8° temperature increase would be close to catastrophic, and result in a 40 percent loss of GDP, again consistent with equation (4.1). The expected value of the temperature increase is still 2° (because $(0.75)(0) + (0.25)(8) = 2°$). But the expected impact of this temperature gamble is now *much greater* than 4 percent of GDP. It is $(0.75)(0) + (0.25)(40) = 10$ percent of GDP. That says that if we had to, we should be willing to sacrifice up to 10 percent of GDP to avoid a 25 percent chance of an 8° temperature increase.

These calculations are summarized in Table 4.1. What's going on here is fairly simple: In terms of its impact on GDP, a 4° temperature increase is more than twice as harmful as a 2° temperature increase. So even though there is only a 50 percent chance of the 4° increase happening, we would sacrifice a lot to avoid the risk. And an 8° temperature increase is more than twice as harmful as a 4° temperature increase, and much more than four times as harmful as a 2° temperature increase. So we would be willing to pay a lot to avoid a very bad outcome, even if that outcome has only a small chance of occurring.

This is the essence of what insurance is all about: We pay to avoid a very bad outcome, even if that outcome is unlikely. That is why we insure our homes against major damage from fire, storms, or floods, why we buy medical insurance to cover the cost of a major hospitalization, and why we buy life

Table 4.1 Possible Temperature Outcomes and Economic Impacts. Impacts are based on the (hypothetical) loss function of equation (4.1), i.e., $L(\Delta T) = 1/(1 + .01(\Delta T)^2)$. Note that in each case the expected temperature change is 2°C, but the expected loss of GDP rises sharply as the maximum possible temperature change rises.

Maximum ΔT Possible	Probability Max ΔT Occurs	Probability of $\Delta T = 0$	percent Loss of GDP if Max ΔT Occurs	Expected Loss of GDP
2°C	1	0	4 percent	4 percent
4°C	0.5	0.5	14 percent	7 percent
8°C	0.25	0.75	40 percent	10 percent

insurance, even if we are healthy and expect to live many more years. And that is why we as a society should be willing to pay a considerable amount for insurance against a very bad (even if unlikely) climate outcome.

GDP Loss and Social Welfare

These simple calculations suggest that we should be willing to sacrifice quite a bit of GDP (and hence quite a bit of consumption) to insure against the risk of a very bad climate outcome. But so far we have only considered one source of insurance value. We focused on the expected loss of GDP (which rises as the maximum possible temperature change rises), but we have implicitly assumed that from the perspective of social welfare, a 10 percent loss of GDP is exactly twice as bad as a 5 percent loss. In fact, a 10 percent loss of GDP might be *more* than twice as bad as a 5 percent loss. The reason has to do with how people value more (or less) income and consumption.

Suppose your annual disposable (after-tax) income is $60,000. Suppose this income is increased to $70,000, so you now have an additional $10,000 to spend on things. That might make you very happy. But now suppose your starting income is $160,000, and we add an extra $10,000, for a total of $170,000. The extra $10,000 will still make you happy, but probably not as much as it would if your starting income was only $60,000. The reason is that with a starting income of $160,000, you already can buy many of the things you want, so the extra $10,000 doesn't give you that much. We call this a *declining marginal utility of income*; the value (in terms of the satisfaction it provides) of an additional $10,000 of income is lower the higher is your starting income.

Of course, for most people climate change will not cause an increase in their income, but instead a reduction. So now let's see what happens if we take away some of your income instead of increasing it. Suppose that climate change damages the economy and as a result causes you to lose some of your disposable income and the amount you can consume.

Let's start with an income of $60,000 and compare a 10 percent loss (i.e., your income drops to $54,000) to a 5 percent loss (your income drops to $57,000). Would the 10 percent loss of income "feel" twice as bad as a 5 percent loss? You can answer this for yourself, but most people would say that a 10 percent loss of income is more than twice as bad as a 5 percent loss. You can see this by taking the 10 percent loss (a loss of $6,000) and dividing it into two 5 percent losses (i.e., $3,000 and another $3,000). The first 5 percent loss reduces your income to $57,000, and that $3,000 reduction hurts. But the second 5 percent loss brings you from $57,000 down to $54,000, and—for most people—that second $3,000 reduction hurts *more* than the first $3,000.

In other words, $3,000 is more valuable if your income is only $54,000 than it is if your income is $57,000. Once again, this is an example of a declining marginal utility of income.

What we call a "declining marginal utility of income" corresponds to *risk aversion*. Let's again consider a reduction in a starting disposable income of $60,000, but this time we'll give you a choice: with Option A, your income will drop by 5 percent (i.e., to $57,000) for certain; with Option B, you flip a coin and if heads, your income stays at $60,000, but if tails, your income drops by 10 percent to $54,000. Which option would you pick? Most people would pick Option A because they are risk-averse; a 10 percent drop in income is more than twice as bad as a 5 percent drop.

One more example: You would probably refuse a lottery in which you had a 50-50 chance of winning $10,000 or losing $10,000. The reason is that (for most people) the value of winning $10,000 is less than the lost value of losing $10,000. How much would you have to be paid to agree to take part in that lottery? Perhaps $2,000, so that you'd have a 50-50 chance of winning $12,000 or losing $10,000? Or perhaps $3,000? The higher the amount you'd have to be paid, the greater is your risk aversion.[6] You can think of this amount you'd have to be paid to take on this risk as an insurance premium.

How risk averse is society as a whole? That question is hard to answer because society is made up of many different people who have very different attitudes towards risk. Financial market data tell us that investors in the aggregate seem to have substantial risk aversion, but not everyone is an investor, and averting climate change is not the same as investing in the stock market.[7]

So what does this tell us about climate policy? It shows us why the uncertainties over climate change are so important, and in particular why society should be willing to sacrifice a substantial amount of GDP to avoid the risk of an extremely bad climate outcome, even if the risk is small. The risk of an extreme outcome—what is sometimes referred to as "tail risk"—might drive us to quickly adopt a stringent emission abatement policy, rather than waiting

---

[6] Economists often describe risk aversion in terms of a *utility function*, which translates income, or wealth or consumption, into units of well-being (or satisfaction). For a textbook explanation, see Pindyck and Rubinfeld (2018). A commonly used utility function is:

$$u(y) = \frac{1}{1-\eta} y^{1-\eta},$$

where y is income and η is called the coefficient of relative risk aversion. In this case marginal utility, i.e., the benefit of an additional dollar of income, is $du/dy = y^{-\eta}$. Marginal utility declines with the level of income, and the larger is η the faster it declines. Thus the larger is η, the greater is the insurance premium you would require to take part in a lottery for which there is a 50-50 chance of winning or losing $10,000.

[7] Based on financial market data, and data on consumption and savings, the coefficient of relative risk aversion (η in the utility function shown the previous footnote) for society as a whole seems to be in the range of 2 to 5, which is substantial.

to see how bad climate change turns out to be. In effect, by reducing emissions now we would be buying insurance. And the value of that insurance could be considerable.

Now, you might be thinking something along the lines of "Well, this is nice. But exactly how large is the value of climate insurance? To what extent does it push us towards early action, and by how much more should we reduce CO_2 emissions if we want to properly account for the insurance value?" Sorry, but I can't provide those numbers, and at this point I don't think anyone can. You may be disappointed with that answer, but keep in mind how little we know. We don't know much about the actual loss function. (The loss function of equation (4.1), which was used to generate Table 4.1, is completely hypothetical.) Nor do we know the extent of risk aversion on the part of society as a whole. All we can say at this point is that the value of insurance is likely to be substantial, and will push policy towards earlier and more stringent emission abatement.

4.1.4 The Effects of Irreversibilities

It has been long understood that environmental damage can be irreversible, which can lead to a more "conservationist" policy than would be optimal otherwise. But we also have to remember that there is a second kind of irreversibility that works in the opposite direction: Protecting the environment imposes sunk costs on society. Keeping our air and water clean imply sunk cost investments in abatement equipment, and an ongoing flow of sunk costs for alternative and perhaps more expensive production processes. If the future value of clean air and water is uncertain, then this kind of irreversibility would lead to policies that are *less* "conservationist" than they would be otherwise. Why? Because if in the future clean air and water turns out to be less valuable than we currently expect, we will regret the irreversible expenditures that were made, and that could have been spent on other things.

Both of these irreversibilities apply to climate policy. Because CO_2 can remain in the atmosphere for centuries, and ecosystem destruction from climate change can be permanent, there is clearly an argument for taking early action. But reducing CO_2 emissions can be quite costly—at least a few percent of GDP—and those costs are largely sunk, which implies an argument for waiting.[8] We know that both of these irreversibilities are important, but

---

[8] There are other arguments for waiting or starting slowly: technological change may reduce abatement costs in the future, and the fact that the "unpolluted" atmosphere is an exhaustible resource implies that the SCC should rise over time (as the atmospheric CO_2 concentration rises).

they work in opposite directions. Which type of irreversibility will dominate depends in part on the nature and extent of the uncertainties involved.

To see the importance of uncertainty, let's focus on the first irreversibility, i.e., over environmental damage. Suppose we know precisely how much environmental damage will result from a higher atmospheric CO_2 concentration, and the damage is completely irreversible. However, we also need to know how society will value that future environmental damage. If we know for certain that the damage will be valued very highly, we would surely want to reduce it by reducing CO_2 emissions now. On the other hand, if we know for certain that the damage won't matter much and won't be valued highly, then we would not devote resources to reducing emissions. The problem is that we *don't know* how those future damages will be valued, and we won't know until the damages actually occur in the future.

If the damage were reversible, then there would be no need to take any action now. If in the future we learn that the damages are highly valued, we would simply "un-do" them (which we could do because we are assuming here that they are reversible). But if they are irreversible, we'd be stuck, and we'd regret not having done more today to reduce emissions. Furthermore, the uncertainty means the damages could end up being valued slightly, moderately, highly, very highly, very, very highly... you get the idea. There is almost no limit to the regret we might feel from not taking action today to reduce emissions and limit those future damages. As a result, the benefit from reducing emissions should include an "option value," which pushes the cost-benefit calculation towards early action.

What about the second irreversibility—the money spent now to reduce emissions is a sunk cost, gone and irretrievable? In this case uncertainty over the value of future damages pushes us to wait rather than take action now. After all, the value of the damages might turn out to be only moderate, or slight, or even zero, in which case we will regret having spent resources today to reduce the damages. Now waiting has an "option value," which pushes the cost-benefit calculation away from early action.

A good way to explore these conflicting irreversibilities, and better understand them, is through the use of a numerical example. In the Appendix to this chapter I provide such an example, which can help to clarify how irreversibilities can interact with uncertainty to affect climate policy.

Emissions Abatement: Hold Back or Accelerate?

This discussion (and the numerical examples in the Appendix) were designed to illustrate the opposing effects of the two irreversibilities that are an inherent aspect of climate policy. But by now you might be asking what we can conclude from this. Which of the two irreversibilities is more important when it comes to actual climate policy? Should we hold back on emissions abatement because of the sunk cost, or should we accelerate abatement because of the irreversible environmental damage caused by emissions? And by how much should we hold back or accelerate?

Unfortunately I can't provide precise answers to these questions. Why not? Because we simply don't know enough about the climate system and about the impact of varying amounts of climate change. To see this, suppose we are considering an aggressive and costly abatement policy that would reduce CO_2 emissions by 80 percent, but we're not sure whether to impose the policy now or wait a few years. In which direction will the two opposing irreversibilities push us? The answer depends in part on how much it would cost to reduce emissions that much, which is unclear. The answer also depends on the impact of the 80 percent emission reduction on future temperature change, and the impact of temperature change on GDP, which we don't know. As a result, while we know that the damages from failing to reduce emissions are highly uncertain, we can't characterize the uncertainty in a way that would let us quantify the effects of the two irreversibilities.

Where does this leave us? The uncertainties over the effects of emission reductions on temperature change and the effects of temperature change on GDP and welfare are so large that we can't determine the net effect of the two opposing irreversibilities. On the other hand, these very large uncertainties imply that the insurance value of early action is large. Whatever the effects of the irreversibilities, they are likely to be swamped by this insurance value—which pushes towards early action.

4.2 Further Readings

The previous chapter explained how and why there is so much uncertainty over climate change; this chapter discussed some of the implications of that uncertainty for climate policy. I emphasized that the considerable uncertainty implies that there is an insurance value to reducing GHG emissions as soon as possible. I also explained that with uncertainty, climate policy is affected by two different irreversibilities that that work in opposite directions: The sunk cost of early emissions abatement versus the ongoing buildup of CO_2 in the atmosphere, which remains there for centuries. There is a growing literature that deals with these issues. Here are some examples:

- For a general discussion of how uncertainty can affect environmental policy (whether or not the environmental problem is climate-related), see Pindyck (2007). Should uncertainty over the extent and impact of climate change lead us to delay abatement? See Litterman (2013) and Pindyck (2013c) for basic arguments as to why the answer is *no*; we should start reducing emissions (with a carbon tax) now.

- It is not just climate policy that is plagued by uncertainty; governments must often make policy decisions in the face of considerable uncertainty. What decision rules should they use to deal with the uncertainty? Manski (2013) provides a thorough and readable treatment of this general problem.

- The implications of uncertainty for climate change policy have also been examined by Heal and Millner (2014), who focus on social welfare aspects of the problem. And some of what is in this chapter also appears in Pindyck (2021).

- One of the earliest studies to analyze the implications of irreversible environmental damage is Arrow and Fisher (1974). As explained in this chapter, there are two different irreversibilities that affect climate policy—the sunk cost of early emissions abatement versus the ongoing and nearly permanent buildup of CO_2 in the atmosphere—and both are important because of uncertainty. A number of studies have explored this question in a theoretical setting; see, e.g., Kolstad (1996), Ulph and Ulph (1997), and Pindyck (2000). These studies illustrate the fundamental problem, but don't go as far as telling us how to formulate climate policy.

- For a thorough textbook treatment of how irreversibilities combined with uncertainty affect economic decisions more generally, see Dixit and Pindyck (1994).

- What can integrated assessment models tell us about the likelihood and severity of a catastrophic climate outcome? Not much, because we lack the theory and data to model extreme outcomes in any convincing way, so the models are essentially hypothetical descriptions of what might happen. On the other hand, a survey of what the models say about a catastrophic outcome gives us a picture of what the modelers think. Kopits, Marten, and Wolverton (2013) provide such a survey.

- Mervyn King, the former Governor of the Bank of England, discusses deep uncertainty (what he calls radical uncertainty) in a different context, but what he says (King (2016), page 131) is very relevant to discussions of climate policy: "The fundamental point about radical uncertainty is that if we don't know what the future might hold, we don't know, and there is no point pretending otherwise." Unfortunately, much of what we read and hear makes it seem as though we know a lot more about climate change and its impact than is actually the case.

4.3 Appendix to Chapter 4: Effects of Irreversibilities

In Section 4.1.4, I explained that there are two kinds of irreversibilities that can affect climate policy (and environmental policy more generally) when there is uncertainty over future costs and benefits. First, environmental damage itself can be irreversible, which can lead to a more "conservationist" policy than would be optimal otherwise. Second, climate policy (e.g., reducing CO_2 emissions) can impose sunk costs on society—sunk cost investments in abatement equipment, and an ongoing flow of sunk costs for alternative production processes. This second kind of irreversibility can lead to policies that are less "conservationist" than they would be otherwise. Both of these irreversibilities are important, but they work in opposite directions, and which one will dominate depends in part on the nature and extent of the uncertainties involved. In this Appendix I use a numerical example to illustrate how this works.

The Example

Suppose we (society) must decide whether to spend money today to reduce CO_2 emissions, and then we will decide whether to spend money again at a point in the future, let's say 40 years from now. We'll assume that at each point in time there are only two choices: (1) Spend nothing on emission abatement ($A = 0$); or (2) spend 6 percent of GDP on abatement ($A = .06$). If today we spend nothing ($A_1 = 0$), there will be 10 units of CO_2 emissions that will accumulate in the atmosphere. So, denoting emissions now by E_1 and the atmospheric concentration by M_1, we will have $E_1 = M_1 = 10$. On the other hand, if we do spend 6 percent of GDP to abate emissions ($A_1 = .06$), emissions will be reduced by 80 percent, so that $E_1 = M_1 = 2$. Finally, we will assume that CO_2 emissions are *partly irreversible*: 50 percent of the today's emissions will dissipate over the next 40 years, so if we emit 10 units of CO_2 today, only 5 units will remain.

To keep this simple, we will also assume that today's emissions cause no damage to the economy now; any damage will occur only in the future. How much damage will occur? We don't know, but suppose there are two possibilities: There is a 50 percent chance that atmospheric CO_2 will cause *no* damage (the "good" outcome) and a 50 percent chance it will cause significant damage (the "bad" outcome). The abatement and outcome possibilities are summarized in Table 4.2.

Suppose there is no abatement now ($A_1 = 0$), so 10 units of CO_2 are emitted. How much abatement would we want in the future? The answer depends on the economic impact, which by then we will know. If the impact

Table 4.2 Example: Immediate Emissions Abatement versus Waiting to Learn about Impact of Warming. A_1 is expenditure on abatement now, as percentage of GDP, and A_2 is expenditure 40 years from now. We denote emissions by E and the amount in the atmosphere by M. If $A_1 = 0$, 10 units of emissions (E_1) will enter the atmosphere (so $M_1 = 10$), but half will dissipate over the next 40 years $(\delta = .5)$. If $A_1 = .06$ (6 percent of GDP is spent on abatement), emissions are reduced by 80 percent, so that $E_1 = M_1 = 2$. Damages occur in 40 years, and depend only on the amount of CO_2 in the atmosphere, $M_2 = (1 - \delta)M_1 + E_2$. The impact is uncertain, and with equal probability could be "good," i.e., no loss of GDP, or "bad," in which case the loss of GDP is $1 - 1/(1 + .03M_2)$, and is shown in the last column. Whatever the value of A_1, if the impact turns out to be "bad," it is best to abate, i.e., set $A_2 = .06$. Also shown is the expected loss of GDP if $A_1 = 0$ (11.5 percent) and if $A_1 = .06$ (7 percent). Since the difference $(11.5 - 7 = 4.5$ percent$)$ is less than the 6 percent cost of abatement, it is better not to abate now, but instead wait and abate in the future only if we learn the impact is "bad."

percent GDP spent on Abatement, A_1	$M_1 = E_1$	percent GDP spent on Abatement, A_2	$M_2 = (1 - \delta)M_1 + E_2$	"Bad Outcome" Loss of GDP
$A_1 = 0$	10	$A_2 = 0$	$5 + 10 = 15$	31 percent
$A_1 = 0$	10	$A_2 = .06$	$5 + 2 = 7$	**17 percent**
Expected Loss if $A_1 = 0$: $(0.5)(0) + (0.5)(.23) = 11.5$ percent				
$A_1 = .06$	2	$A_2 = 0$	$1 + 10 = 11$	25 percent
$A_1 = .06$	2	$A_2 = .06$	$1 + 2 = 3$	**8 percent**
Expected Loss if $A_1 = .06$: $(0.5)(0) + (0.5)(.14) = 7$ percent				

is zero (the "good" outcome), there would be no reason to abate, so we will have $A_2 = 0$. (This outcome is not shown in the table.) But if the "bad" outcome occurs (an 8 percent loss of GDP), we will want to abate emissions, i.e., set $A_2 = .06$. As Table 4.2 shows, with the "bad" outcome and $A_2 = 0$, the loss of GDP will be 31 percent, but with $A_2 = .06$, the loss of GDP will only be 17 percent. Abatement will cost 6 percent of GDP, but we will save $(31 \text{ percent} - 17 \text{ percent}) = 14$ percent of GDP, so the investment in abatement is clearly worth it.

Why not set $A_1 = .06$ at the outset, before we learn whether the impact will be "bad" or "good"? The reason is that spending 6 percent of GDP on abatement is an irreversible expenditure which we will regret if it turns out the impact of climate change is "good." But to know whether the potential regret is large enough, we have to see what happens if we *do* set $A_1 = .06$ at the outset. As Table 4.2 shows, with $A_1 = .06$, only 2 units of CO_2 will be emitted, and of those 2 units, only 1 will remain after 40 years. If we then learn that the impact of climate change is "good," there will be no reasons to abate, so we will set $A_2 = 0$. But if the impact is "bad," it will be best to abate, so we will set $A_2 = .06$ (which yields a loss of GDP of 8 percent, versus a 25 percent loss if we set $A_2 = 0$).

Now let's come back to the initial decision regarding A_1. What is the expected loss of GDP if we set $A_1 = 0$? As shown in Table 4.2, there is a 50 percent chance that the impact will turn out to be "bad," in which case we will set $A_2 = .06$ (which costs 6 percent of GDP), but still lose 17 percent of GDP (because $1 - 1/(1 + .03 \times 7) = .17$), for a total GDP loss of 17 percent $+ 6$ percent $= 23$ percent. So the expected loss if $A_1 = 0$ is $(0.5)(0) + (0.5)(23$ percent$) = 11.5$ percent. Also shown is the expected loss of GDP if $A_1 = .06$, which turns out to be 7 percent. Since the difference $(11.5 - 7 = 4.5$ percent$)$ is less than the 6 percent cost of abatement, it is better *not to abate now*, but instead to wait and abate in the future only if we learn the impact is "bad."

To summarize, we have assumed that CO_2 emissions are only partly irreversible, i.e., 50 percent of the today's emissions will dissipate over the next 40 years. The cost of abatement (6 percent of GDP), however, is completely irreversible; it is a sunk cost that can never be recovered. In this case, given that the impact of CO_2 is uncertain and will only be known in the future, it is better to wait, rather than spend 6 percent of GDP now on abatement. Because half of today's emissions will dissipate, the abatement cost irreversibility outweighs the environmental irreversibility.

Revising the Example

But now let's change one of the key assumptions and then repeat these calculations. This time we will assume that there is *no dissipation* of CO_2 once it enters the atmosphere, so the environmental damage is completely irreversible. This means setting $\delta = 0$ so that $M_2 = M_1 + E_2$. The results are shown in Table 4.3.

Because we have now assumed that any CO_2 emitted into the atmosphere stays there forever, the loss of GDP under the "bad" outcome will be greater, whatever the abatement policy happens to be. (Compare the last column of Table 4.3 with the last column of Table 4.2.) As in the previous example, whatever the value of A_1, if in the future the impact turns out to be "bad," it is best to abate, i.e., set $A_2 = .06$.

What is the optimal abatement policy today, before we know whether the impact will be "good" (no impact) or "bad"? As in the original example, we find out by calculating the expected loss of GDP if we set $A_1 = 0$, and the expected loss if we set $A_1 = .06$. If $A_1 = 0$ the expected loss of GDP is

16 percent, and if $A_1 = .06$ the expected loss is 8.5 percent. Now the difference $(16 - 8.5 = 7.5$ percent$)$ is greater than the 6 percent cost of abatement, so it optimal to abate immediately. Because emissions are now completely irreversible (there is no dissipation), we are pushed towards early action. The sunk (irreversible) cost of abatement remains, pushing us towards waiting, but now the effect of the environmental irreversibility dominates.

A Homework Exercise

As a homework exercise, try the following. Modify the numbers in Table 4.3 with one simple change: Assume that positive abatement requires an expenditure of 8 percent of GDP rather than 6 percent. (We will still assume that there is no dissipation, i.e., whatever CO_2 is emitted in the beginning will remain in the atmosphere over the 40 years.) So in Table 4.3, replace $A_1 = .06$ with $A_1 = .08$ and $A_2 = .06$ with $A_2 = .08$. Now you can calculate the expected loss if $A_1 = 0$ and the expected loss if $A_1 = .08$. If you go through this exercise, you will find that the expected loss is 17 percent if $A_1 = 0$ and 9 percent if $A_1 = .08$. The difference, $17 - 9 = 8$ percent is just equal to the 8 percent cost of abatement. In this case the effects of the two irreversibilities just balance out, so we would be indifferent between abating now and not abating now (and we could flip a coin to decide).

Table 4.3 Modified Example: Immediate Emissions Abatement versus Waiting for Information. Everything here is the same as in Table 4.2, except the dissipation rate, δ, is zero, so that emissions are completely irreversible. Whatever the value of A_1, if in the future the impact turns out to be "bad," it is best to abate, i.e., set $A_2 = .06$. Also shown is the expected loss of GDP if $A_1 = 0$ (16 percent) and if $A_1 = .06$ (8.5 percent). Now the difference $(16 - 8.5 = 7.5$ percent$)$ is greater than the 6 percent cost of early abatement, so it optimal to abate immediately. Because emissions are now completely irreversible (there is no dissipation), we are pushed towards early action. The sunk (irreversible) cost of abatement remains, pushing us towards waiting, but now the environmental irreversibility dominates.

percent GDP spent on Abatement, A_1	$M_1 = E_1$	percent GDP spent on Abatement, A_2	$M_2 = (1 - \delta)M_1 + E_2$	"Bad Outcome" Loss of GDP
$A_1 = 0$	10	$A_2 = 0$	$10 + 10 = 20$	37.5 percent
$A_1 = 0$	10	$A_2 = .06$	$10 + 2 = 12$	**26.5 percent**
Expected Loss if $A_1 = 0$: $(0.5)(0) + (0.5)(.325) = 16$ percent				
$A_1 = .06$	2	$A_2 = 0$	$2 + 10 = 12$	26.5 percent
$A_1 = .06$	2	$A_2 = .06$	$2 + 2 = 4$	**11 percent**
Expected Loss if $A_1 = .06$: $(0.5)(0) + (0.5)(.17) = 8.5$ percent				

5

Climate Policy and Climate Change: What Can We Expect?

In Chapter 2 we used an extremely simple model to examine the likelihood of preventing the increase in the global mean temperature from exceeding 2°C by the end of the century. We saw that even if we assume an optimistic scenario for future CO_2 emissions (they decline to zero between 2020 and 2100), it is likely that the temperature change will be well above 2°C. There is uncertainty, of course, particularly over the value of climate sensitivity. We used a value of 3.0 for climate sensitivity, which is in middle of the "most likely" range. If the actual value of climate sensitivity turns out to be significantly lower, the temperature change could well stay below 2°C. But if climate sensitivity turns out to be higher than expected, the temperature change could go as high as 4°C. Put simply, if we were to bet on a temperature change below 2°C (and lived long enough to collect on the bet), the odds would be against us.

This chapter explores what might or might not happen in more detail. We start with CO_2 emissions and ask what kinds of scenarios might we view as realistic, whether or not they are optimistic. Given that some countries (and some parts of the U.S.) have passed laws requiring (or at least targeting) net-zero CO_2 emissions by 2050, can't we do better than the simple optimistic scenario we examined in Chapter 2? The U.S. has re-joined the Paris Agreement, and if the Agreement is revised and extended during the coming years to require broader and more rapid emission reductions, can't we push global CO_2 emissions to zero well before the end of the century? Perhaps. After all, the uncertainties over the future climate policies that different countries will adopt are about as great as the uncertainties over the climate system itself. But we will see that for the world as a whole, although reaching zero emissions well before the end of the century is possible, it is unlikely, and certainly not something that we should count on.

Our discussion so far has ignored methane, except for my claim on page 29 that it makes a much smaller contribution to climate change than does CO_2. A ton of methane, however, is about 28 times as powerful as a ton of

Climate Future: Averting and Adapting to Climate Change. Robert S. Pindyck, Oxford University Press.
© Oxford University Press 2022. DOI: 10.1093/oso/9780197647349.003.0005

CO_2 in terms of its warming potential, so how can we say that it doesn't contribute as much to climate change? The answer is that far fewer tons of methane than CO_2 are emitted each year, and methane only stays in the atmosphere for a decade or so, while CO_2 stays there for centuries. As a result, methane accounts for less than 20 percent of the total warming effects of GHG emissions. On the other hand, 20 percent is much more than zero, so we will take some time to look at methane in more detail, and evaluate its potential contribution to climate change over the coming decades.

Given alternative scenarios for CO_2 emissions (some realistic, some less so), and some alternative scenarios for methane emissions, we will examine the possible implications for changes in the global mean temperature. It is important to stress the word "possible," because given all of the uncertainty over climate sensitivity (and other aspects of the climate system), the best we can do is come up with a range of possible outcomes for any particular emission scenario. Looking at ranges of possible outcomes is useful, however, because it will help clarify the risk that society faces, and the value of "climate insurance" to reduce that risk.

5.1 CO_2 Emission Reductions

I argued at the outset of this book that world GHG emissions are likely to keep growing, at least over the coming decade. Yes, the U.S. and Europe have already made significant progress in reducing emissions, and are likely to make more progress. But what matters for climate change is *global* emissions. Emissions from Asian countries, most notably China, India, Malaysia, and Indonesia, are large and have been growing rapidly. Emissions from Latin America, Africa, and the Middle East are currently more modest, but as with Asia, are growing rapidly. Why is this rapid growth in emissions occurring? The main reason is economic growth. In the past, the economies of these countries were far less developed. But as their economies began to grow rapidly, their emissions likewise grew. And those increased emissions have completely swamped the (relatively small) emission reductions in the U.S. and Europe.

Furthermore, on a per capita basis, CO_2 emissions in most of Asia (as well as Africa and Latin America) are still well below levels in the U.S. and Europe. This can be seen in Figure 5.1, which shows total emissions and per capita emissions for the 15 countries that were the largest emitters of CO_2 in 2017. Note that China had by far the largest CO_2 emissions, almost double that of the U.S. But China's per capita CO_2 emissions were only about half as large

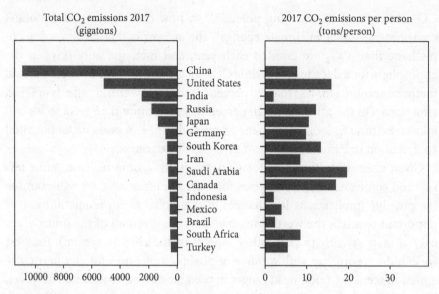

Fig. 5.1 Total and Per Capital CO_2 Emissions in 2017.
Source: Emissions Database for Global Atmospheric Research (EDGAR), 2018 Report.

as the U.S. The story is similar for India; its total CO_2 emissions were about half that of the U.S., but its per capita emissions were less than one-eighth of the U.S.

Why be concerned with per capita emissions? Because reducing emissions is costly. So a relatively poor country like India that has far less per capita emissions than the U.S., Europe, Japan, Russia, and many other wealthier countries would naturally object to being asked to make large percentage reductions in their total emissions. India would argue (as it already has) that it should not have to bear anywhere near the burden that a wealthy country like the U.S. should have to bear. But that makes it very difficult to reach an international agreement that pushes *all* countries to sharply reduce their CO_2 emissions. And an international agreement—one that is enforceable—is crucial if we want to reduce global emissions. Without such an agreement, it is just not realistic to think that global CO_2 emissions will suddenly stop rising and instead gradually fall to zero by the end of the century, along the lines of the scenario shown in Figure 2.1.

Let's look at the prospects for CO_2 emission reductions over the coming decades. We'll begin with the bright side—the U.S. and Europe, which have already succeeded in reducing emissions significantly, and are likely to continue reducing emissions. (Brexit aside, I am including the U.K. with Europe.) Then we'll look at other parts of the world, where the picture is much less bright. The most important country in this regard is China, which has become the largest emitter of CO_2 and other GHGs, where emissions are continuing to grow, and where the near-term potential for substantially reducing emissions is bleak. Finally, we will look at India and other Asian countries, as well as countries in Latin America and other parts of the world. Here, too, we will see that the outlook is not very good.

Again, if only the U.S. and Europe produced GHGs, the outlook for substantial emission reductions during the next two or three decades would be much, much better. By 2019 (prior to the further drop in emissions caused by the COVID pandemic), CO_2 emissions in the U.S. had fallen about 14 percent from their peak in 2007, and emissions from the U.K. and many European countries had fallen even more. What was the cause of these declines, and can we expect them to continue?

5.1.1 The United States

Let's begin with the U.S. Almost all CO_2 emissions come from energy consumption of one form or another—burning coal or natural gas to produce electricity, burning oil and natural gas to heat homes and factories, and burning gasoline and jet fuel for transportation. Figure 5.2 shows CO_2 emissions from energy consumption in the U.S. from 1975 onwards. Apart from the drop during the period 1979 to 1982, emissions rose steadily until 2007,

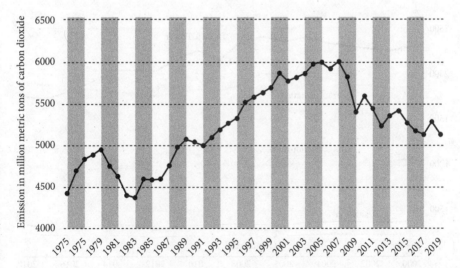

Fig. 5.2 CO_2 Emissions from Energy Consumption in the U.S., millions of metric tons per year.

Source: U.S. Energy Information Agency, *Monthly Energy Review*, 1975 to 2019.

peaking just before the financial crisis in late 2007 and Great Recession that began in 2008.

Why the drop from 1979 to 1982? Three reasons. First, that was the period of the Iranian Revolution and the Iran-Iraq War, which sharply reduced oil production from Iran and Iraq, pushing up oil (and therefore gasoline) prices, and thereby reducing demand. Second, the U.S. imposed price controls on oil, which created gasoline shortages, reducing gasoline consumption. And third, the U.S. economy experienced recessions in 1980 and 1982, which drove down energy consumption and thus CO_2 emissions.

And why the steady increase in CO_2 emissions (at an average rate of 1.3 percent per year) from 4.4 Gt in 1983 to 6.0 Gt in 2007? The main reason is that the U.S. economy grew steadily during those 24 years, and with little in the way of conservation, this pushed up energy consumption. But after 2007, emissions started to fall as a result of the financial crisis and the Great Recession. During the worst of the Great Recession (2008 to 2010), energy consumption fell sharply, as did CO_2 emissions.

Starting around 2010 the Great Recession largely ended and economic growth began to pick up. Yet CO_2 emissions continued to fall, at a rate of around 1 percent per year. Why did that happen, and most important, can it continue? The drop in emissions after 2010 is due to a combination of factors, but the main one is the drop in coal consumption. To see this, we have to look at electric power generation and its role in CO_2 emissions.

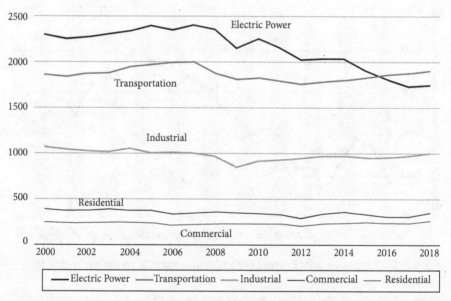

Fig. 5.3 Energy-Related U.S. CO_2 Emissions by Sector, millions of metric tons per year. *Source*: U.S. Energy Information Agency, *Monthly Energy Review*, 2019.

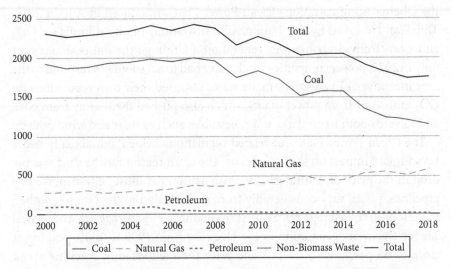

Fig. 5.4 U.S. CO_2 Emissions from Electric Power Generation by Fuel, millions of metric tons per year.
Source: U.S. Energy Information Agency, *Monthly Energy Review*, 2019.

Figure 5.3 shows how the decline in U.S. CO_2 emissions was driven by the electric power sector. Note that until 2016 electric power generation was the largest source of CO_2 emissions in the U.S. Emissions generated from the transportation sector (mostly automobiles, but also trucks, buses, and air travel) were about the same in 2018 as in 2000, but as emissions from electric power fell, transportation became the largest source of U.S. emissions. Emissions from industrial, residential, and commercial use of energy were also about the same in 2018 as in 2000. But annual emissions from electric power generation have dropped substantially from 2006 (about 2.4 Gt) to 2018 (about 1.7 Gt).

What caused the drop in emissions from electric power? It's not that the U.S. is consuming less electricity than it did a decade ago, but rather it's the way the electricity is made. In the past, most electricity was made by burning coal. Apart from other harmful emissions (particulates, sulfur dioxide, and nitrogen oxides), burning coal produces large amounts of CO_2—about twice as much (per BTU of energy produced) compared to burning natural gas. Starting around 2008, the U.S. shifted rapidly from coal to natural gas, increasing the carbon efficiency of electricity production. This shift can be seen in Figure 5.4, which shows CO_2 emissions from electric power generation by fuel type.

Why did electricity production move from coal to natural gas? To some extent the shift was driven by environmental regulations, and in particular the Obama Administration's Clean Power Plan, announced in August 2015. This Plan (reversed by the Trump Administration) imposed limits on CO_2 emissions from electric power generation. (Although the limits would start only in 2022, power companies would have had to act sooner to respond to the new incentives.) In addition, many states imposed their own regulations on CO_2 emissions from power plants, which also pushed them away from coal, and towards both natural gas and renewables such as solar and wind power.

The Clean Power Plan and related regulations helped, but actually had a very limited impact on the use of coal. The main reason for the shift was the drop in the price of natural gas. Because natural gas must be transported by pipelines, prices vary considerably from one location in the U.S. to another, so we look at the average price across the country. The average price of natural gas for electricity generation in the U.S. rose from 1995 to 2005, peaking at close to $8 per million BTUs.[1] The price then fell steadily, reaching about $3.42 per MMBtu in in 2012, and less than $3.00 in 2019.

The decline in the price of natural gas was largely driven by the advent and rapid growth of fracking in the U.S., making the fuel plentiful and cheap— but not cheaper on a BTU basis than coal. In 2019–2020, the average prices of natural gas and coal for electricity generation were both around $2 to $3 per MMBtu. But this apparent price equivalence actually made natural gas much more economical as a fuel. The reason is that coal is dirty. Apart from CO_2 emissions, burning coal produces emissions of particulates (atmospheric particles that can be made up of acids, organic chemicals, and soil or dust), sulfur dioxide (SO_2), and nitrogen oxides (nitrous oxide, NO, and nitric oxide, NO_2). All of these emissions can travel many miles from their source, and pose substantial health hazards. Thus they are regulated at both the federal level (through the Clean Air Act) and by individual states. To burn coal, power plants have to install expensive equipment ("scrubbers") to prevent most of these emissions from entering the atmosphere or, alternatively, burn more expensive low-sulfur coal. The higher capital and operating costs of a coal-burning plant makes natural gas the more economical choice for a new power plant, even if the price of gas is somewhat higher than the price of coal.

To sum up, CO_2 emissions from the U.S. have been declining since 2007, driven largely by the shift away from coal in electricity generation. But what will happen to U.S. emissions over the coming decades? If the U.S. imposes a stiff carbon tax and adopts other policies along the lines of the "Green New Deal," can the decline in emissions continue and even accelerate? What is the most we can hope for?

---

[1] To meaningfully compare fuel prices we usually express them in terms of dollars per quantity of energy produced from burning the fuel, typically dollars per million BTUs ($/MMBtu). The energy content of 1000 cubic feet (1 mcf) of natural gas is roughly 1 MMBtu.

A stiff carbon tax—something on the order of $100 per metric ton of CO_2—would have the greatest impact on emissions. It would raise the price of gasoline by around $1.00 to $1.50 per gallon, raising the demand for more fuel-efficient cars, and thereby reducing gasoline consumption in the long run (after 5 to 8 years) by some 20 to 30 percent. By raising the prices of both coal and natural gas, a carbon tax would accelerate the adoption of solar and wind power in electricity generation, and would also reduce electricity consumption. And it would significantly reduce industrial consumption of oil and natural gas.

A carbon tax is efficient, effective, and simple—and in the U.S., at least, extremely unpopular. Perhaps it will become more popular, or at least politically feasible, in the future. But for at least the next several years it is not something we are likely to see. Instead, climate policy is more likely to consist of direct or indirect regulations on energy use. An example is corporate average fuel economy (CAFE) standards that impose a gasoline mileage standard (e.g., 35 miles per gallon) on the fleet of cars and light trucks that each auto maker sells. Another example is a requirement forcing electricity producers to use more renewables instead of fossil fuels. And climate policy is also likely to include subsidies, both for the use of "green" energy (mainly solar and wind), but also for R&D into new technologies that would reduce fossil fuel use.

These sorts of policies will certainly help reduce CO_2 emissions (although the impact of R&D subsidies is highly uncertain and could take many years). But it is unlikely that these policies alone will lead to the elimination of fossil fuel use by, say, 2050. The problem is the highly political nature of climate policy. For example, on January 10, 2019, a letter signed by 626 organizations in support of a Green New Deal, sent to all members of Congress, called for measures such as "a ban on crude oil exports; an end to fossil fuel subsidies and fossil fuel leasing; and a phase-out of all gasoline-powered vehicles by 2040." But the letter also said signatories would "vigorously oppose . . . market-based mechanisms and technology options such as carbon and emissions trading and offsets, carbon capture and storage, nuclear power, waste-to-energy and biomass energy." Tying our hands in this way is a natural political outcome, but greatly reduces the chances that the U.S. will reduce CO_2 emissions to anything close to zero by mid-century.

5.1.2 The U.K. and Europe

By 2020, the U.K. and Europe had made considerably more progress than the U.S. in reducing emissions, and most of the gains were policy-driven. Let's look first at the U.K., which by 2019 had reduced its CO_2 emissions by about 45 percent below 1990 levels, more than most other countries.

The U.K.

The UK's Climate Change Act (CCA) of 2008 set a goal of an 80 percent emissions reduction relative to 1990 levels by 2050. This is an ambitious goal, but in June 2019, the U.K. parliament decided it wasn't ambitious enough, and revised the goal to *zero net emissions* by 2050. Although the U.K. has made considerable progress, it is unclear whether even the earlier goal of an 80 percent reduction can be met, never mind the net-zero goal. (Each goal was supposed to be legally binding, but it's unclear what would happen should the goal not be met. Would any politicians go to jail? Probably not, but we'll see.)

As with most countries, the U.K.'s emission reductions achieved to date have been almost entirely in the generation of electricity, which in 1990 was heavily dependent on coal and accounted for some 40 percent of total emissions. Emission reductions in electricity generation have resulted from phasing out coal and increasing the use of renewables. The problem is that coal's share of the U.K.'s electricity production is now so low that there is very limited scope to continue reducing emissions by reducing the use of the fuel. (As of 2020, only four coal-fired power plants were operating in the U.K., one of which was due to be retired in 2021, and the other three a year or two later.) As a result, further emission reductions will have to come from other sectors of the U.K. economy, such as transportation, home heating, and industrial production. But unfortunately there has been little or no reduction in emissions in these other sectors, and it is not clear where any significant reductions might come from. That's why the U.K. is not on target to meet even the 80 percent goal.

With the gains made in electricity production, the policy focus in the U.K. will turn to transportation, which is now the sector producing the greatest amount of emissions. In early 2020, the U.K. government announced that there will be a ban on the sale of new gasoline-powered cars (including diesel and hybrid cars) by 2035. If this ban is enforced, by 2040 most cars will be electric, and if most of the electricity can be generated by renewables, this could reduce emissions by 10 or even 15 percent. But enforcing a ban on gasoline-powered cars will not be easy; electric vehicles need charging stations, which will be difficult to install in the winding streets of most U.K. cities. Reducing natural gas consumption will also be difficult; around 85 percent of homes in the U.K. are heated by gas, which for heating is more efficient than electricity, and limitations on the electricity grid makes it unlikely that any significant fraction of those homes will be converted to electricity.

The bottom line: It is likely that the U.K. will continue to reduce its GHG emissions, but it is unlikely that it will meet its net-zero (or even 80 percent) target for 2050. Of course "unlikely" is not the same as impossible, and many in the U.K. are hopeful that the net-zero target will indeed be met.

Europe

Now let's move to Europe. Emissions have dropped substantially for the European Union as a whole, although not by as much as the U.K. From 1990 to 2018, total GHG emissions from the EU fell by 21 percent. This puts the EU half way towards its target for 2030—a 40 percent reduction from 1990 levels. That 40 percent target was set by the European Council in October 2014. Although it has been looking increasing unlikely that the 40 percent target could be met, in 2020 the European Commission proposed a "European Green Deal" that set stricter targets: It raised the 2030 target to a 50 percent reduction, and set a target for 2050 of zero net emissions, matching the U.K. target for 2050.

But again, the problem is targets versus outcomes. What happens if the EU fails to meet its 2030 target? Most likely nothing, except for a pledge to meet the 40 percent target—or even a more stringent target—as soon as possible. There are two problems here. First, reducing emissions by 40 percent is difficult, more than twice as difficult as reducing emissions by 20 percent. Second, there is considerable heterogeneity across European countries in terms of their energy use and steps they have taken or could take to reduce emissions. Some European countries will meet the 40 percent target relatively easily, but for other countries it will be much more difficult.

The heterogeneity is illustrated by Figure 5.5, which shows CO_2 emissions from four of the European Union Countries, along with the U.K., over the period 2005 to 2019. In terms of total emissions, Germany is by far the leader; its CO_2 emissions averaged around 750 million metric tons (0.75 Gt) during this period, compared to around 350 for France. On a per capita basis, that amounts to 9.0 tons per person for Germany (population 83 million in 2020) compared to 5.2 tons per person for France (population 67 million), 5.8 tons per person for Italy, 7.9 for Poland, and 7.1 for the U.K. Germans emit more CO_2 per person than most other European countries because they consume more energy per person, mostly for transportation, but also for electricity generation. And much of this energy comes from coal. (In 2019, Germany ranked 4th in the world for coal consumption.)

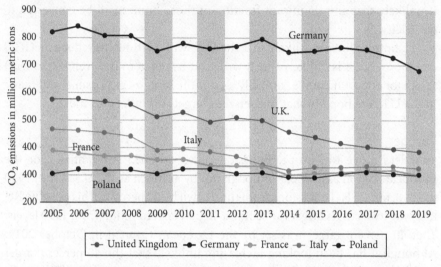

Fig. 5.5 CO_2 Emissions from the U.K. and Four European Union Countries, 2005–2019, in million metric tons.
Source: BP and Statistica 2020.

In 2005 Germany obtained about 30 percent of its electricity from 17 operating nuclear power plants. But because of strong public opposition to nuclear power, the German government decided in 2011 to eliminate it. By 2019, 10 of the 17 nuclear plants had been shut down, there are no plans to build any new nuclear plants, and nuclear power only accounted for 11 percent of electricity production. This made it harder for Germany to reduce its dependence on coal. (Germany announced in January 2020 that it would spend $44.5 billion to completely eliminate the use of coal—but not until 2038.) On the other hand, Germany has made substantial progress in the shift to renewables (mostly wind, but also solar and biomass), which by 2019 accounted for about 40 percent of its electricity generation.

The heterogeneity across EU members also applies to the progress (or lack thereof) that countries have made in reducing CO_2 emissions. Note from Figure 5.5 that Poland has made no progress over this period; its emissions in 2019 were almost exactly the same as in 2005. Germany reduced its emissions by about 16 percent, France by about 24 percent, and both Italy and the U.K. by about 33 percent. Some countries will be able to meet the EU target of a 40 percent reduction in emissions by 2030, but some (certainly Poland, which at this point is wedded to coal, and probably Germany, given its abandonment of nuclear power) will not.

5.1.3 China

In 2020 China accounted for close to 30 percent of the world's CO_2 emissions, which was roughly double that of the U.S. (about 10 Gt versus 5.5 Gt for the U.S.). And from 2018 to 2019, China's CO_2 emissions continued to grow (by about 2.5 percent), even though emissions from the rest of the world were flat during this period.[2]

So how difficult would it be for China to substantially reduce its emissions, and (putting the COVID pandemic aside) what should we expect? In a message to the UN General Assembly in September 2020, China's leader Xi Jinping made a pledge: China's CO_2 emissions would stop increasing by 2030, and by 2060 its net emissions would fall to zero. That pledge is both dramatic and encouraging, but as with Europe, the U.S., and other countries, pledges and outcomes can differ considerably. There is certainly room for China to reduce its emissions, but doing so will not be easy, and for the most part the reductions will have to be part of an international climate agreement.

What are the difficulties? First, remember that on a per capita basis, China's CO_2 emissions were less than half of U.S. (7.1 tons per person, versus 16.6 for the U.S.). Thus China (along with other countries with low per capita emissions, such as India) would naturally object to making large reductions in their total emissions. Those countries would argue that they should not have to bear anywhere near the burden that the U.S. and other wealthy countries should have to bear.

Second, China's heavy dependence on coal will not be easy to reverse. Yes, even putting climate change aside, China has incentives to reduce coal consumption; the main incentive being to reduce the high levels of air pollution (mostly particulates) that plague many of the country's cities. But China's economic growth implies substantial growth in its demand for electricity, and coal provides the cheapest way to produce that electricity. China's electric generating capacity has grown by about 6 percent per year over the past few years, and in 2019 about 60 percent of its electricity was produced by coal. Many new coal-fired power plants are under construction in China, and more are in the planning stages.

Third, China's economy remains one of the fastest-growing in the world. Using exchange rates to convert the yuan to U.S. dollars, China's GDP in 2020 was about $14 trillion, compared to $21 trillion for the U.S. But converting via

---

[2] Global emissions fell sharply during the first half of 2020 because of the COVID pandemic, which shut down much of transportation and industrial production around the world. As the pandemic ends and world GDP starts to returning to normal, CO_2 emissions are returning to the pre-COVID growth path.

exchange rates is not the most informative way to compare economies, and economists often use a Purchasing Power Parity (PPP) Index instead.[3] Using a PPP Index instead of an exchange rate, China's 2020 GDP was around $27 trillion, larger than the U.S. GDP. With the world's largest GDP (measured in PPP terms) and rapidly growing economy, without a major policy shift China's CO_2 emissions could continue growing until at least 2030.[4]

Trade-Adjusted Emissions

Before moving to the rest of the world, it is worth noting that the numbers for CO_2 emissions can be adjusted to account for international trade. CO_2 emissions are typically measured on a territorial basis, i.e., the amount emitted within the geographical boundaries of the country—even if some of the emissions resulted from the production of goods that the country exported. As a result, the emission numbers are "production-based." However, statisticians also calculate "consumption-based" emissions, which are adjusted for trade.

To calculate consumption-based emissions we track which goods were traded across the world, and if a good was imported we include the CO_2 emissions that were emitted in the production of that good, and subtract the emissions that came from the production of goods that were exported. For example, suppose that 1 Gt of China's CO_2 emissions came from its production of consumer electronics that were exported to and consumed in the U.S. Then we would reduce China's consumption-based emissions by 1 Gt and increase the U.S. consumption-based emissions by 1 Gt. (Total emissions, of course, remain the same; we are simply reallocating the sources of those emissions.)

Why is this important? Because to reach an international agreement, countries negotiate over their emissions reductions, and each country wants other countries bear the burden of reducing emissions. If a country's consumption-based CO_2 emissions are lower than its production-based emissions, it can argue that it shouldn't have to reduce its emissions so much because other countries (with higher consumption-based emissions) are benefiting from its production. We can see this by looking at the U.S. and China.

---

[3] Exchange rates are determined by flows of traded goods and flows of capital, but much of what people consume is not traded (e.g., housing, transportation, and food) and people do not (directly) consume capital. Unlike an exchange rate, a Purchasing Power Parity (PPP) Index allows us to convert currencies from one country to another in terms of what people actually consume in the two countries.

[4] Forecasting is speculative and the forecasts vary, but see: https://climateactiontracker.org/countries/china.

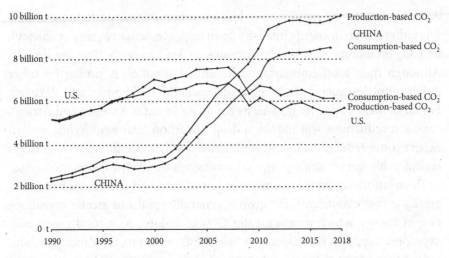

Fig. 5.6 CO_2 Emissions from the U.S. and China, on a Production versus Consumption Basis.

Source: www.globalcarbonproject.org, and see the discussion in Peters, Davis, and Andrew (2012).

Consumption-based versus production-based CO_2 emissions for China and the U.S are shown in Figure 5.6. China exports much of its production, and the U.S. has been a net importer. So, as you would expect, China's consumption-based CO_2 emissions are nearly 20 percent lower than its production-based emissions, whereas for the U.S., the consumption-based emissions are about 10 to 15 percent higher. When negotiating emission reductions, China could reasonably argue that a good part of its emissions are actually for the benefit of consumers in the U.S. (and other countries that import Chinese goods).

5.1.4 The Global Picture

China is not the only country with large and growing levels of CO_2 emissions. As Figure 1.2 shows, except for Europe, the U.K., and the U.S., emissions have been growing steadily in the rest of the world. And for many countries, the difficulty in reversing that growth is the same as is for China—their economies are growing and government policy has prioritized continuing growth and the alleviation of poverty over GHG emission reductions.

India is a good example of the problem. In 2018 its total CO_2 emissions were about 2.7 Gt, but its per capita emissions were only about 2 tons per person, which is one-eighth of the per capita emissions of the U.S. India's economy is growing rapidly, so unless its carbon intensity falls substantially (via either a drop in energy intensity or an improvement in energy efficiency), its CO_2 emissions are likely to increase, at least over the coming decade. Although their total emissions are lower, the situation is similar for other Asian countries, such as Indonesia, Malaysia, the Philippines, and Pakistan. For all of these countries (as well as countries in Latin America and Africa), emission reductions will require a drop in carbon intensity.[5] While we can expect some reductions in carbon intensity, unless economic growth slows considerably (an unhappy prospect), emissions are likely to continue to rise.

The relationship between economic growth and CO_2 emissions has varied greatly across countries. GDP growth generally results in greater consumption of energy, which means greater CO_2 emissions. As a result, we would expect per capita CO_2 emissions to increase as per capita GDP increases, and indeed that has been the case. Figure 5.7 plots per capita CO_2 emissions (in

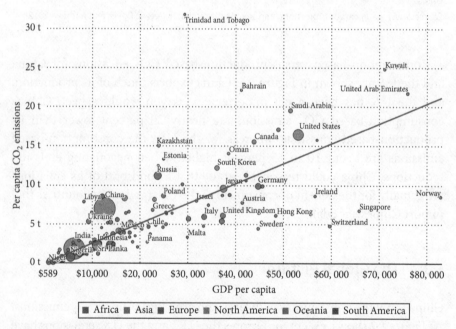

Fig. 5.7 Per Capita CO_2 Emissions versus Per Capita GDP. Annual CO_2 emissions (vertical axis) are measured in metric tons per person, and per capita GDP (horizontal axis) is measured in 2011 (inflation-adjusted) U.S. dollars. Data are for the year 2016, and each country is shown as a circle with area proportional to that country's total CO_2 emissions. The straight line is a best fit to the data.
Source: Global Carbon Project, OurWorldinData.org; best fit line added by the author.

[5] If you've forgotten the meaning of carbon intensity, energy intensity, and carbon efficiency, go back to page 45, where these concepts are explained.

metric tons) against per capita GDP (in 2011 U.S. dollars) for a large group of countries. (The data are for 2016, and each country is shown as a circle, with area proportional to that country's total CO_2 emissions.) As you'd expect, those countries with greater per capita GDP also tended to have greater per capita emissions.

But note that the relationship between per capita CO_2 emissions and per capita GDP is far from perfect. Also shown on the graph is straight line that has been fit to the data. If the relationship between per capita CO_2 emissions and per capita GDP were exact, all of the circles would lie on that straight line. But they don't. China's per capita emissions of 7.1 tons, for example, is more than double the roughly 3-ton level predicted by its per capita GDP. Middle Eastern countries (e.g., Saudi Arabia and Kuwait) also have per capita emissions well above the fitted line. On the other hand, the per capita emissions of the U.K. and most European countries are well below the levels predicted by their per capita GDPs. Switzerland, for example, emitted 5 tons of CO_2 compared to the 15 tons predicted by its per capita GDP.

Figure 5.7 shows that countries' CO_2 emissions are not completely constrained by their economic output or rate of economic growth, and carbon intensity can be reduced. There is room to reduce energy intensity via policies (or price incentives) that increase the fuel efficiency of cars, and improve the efficiency of heating, cooling, and refrigeration. Likewise, there is room to improve energy efficiency, i.e., reduce the amount of CO_2 emitted from each quad of energy. At issue is whether the reductions in carbon intensity will be extensive enough and come soon enough to cause global emissions to begin declining, and then continue declining towards zero. If not, a 2°C limit is not something we should count on.

Can a Pandemic Save Us?

During 2020, there was a significant decline in CO_2 emissions. No surprise— the COVID-19 pandemic prevented people from traveling, attending sports and cultural events, and in many cases even going to work. The result was large drop in energy consumption, especially the consumption of gasoline and jet fuel, and a corresponding drop in CO_2 emissions.

But the drop in emissions was temporary. As the pandemic started to come under control and life gradually started to return to normal, emissions likewise started to return to "normal," which means high and increasing over time. In fact, the data for 2020 make this abundantly clear. Figure 5.8, which is an updated version (by the Global Carbon Project and United Nations Environment Programme (2020)) of the graph by Le Quéré et al. (2020),

Change in global daily fossil CO_2 emissions, percent

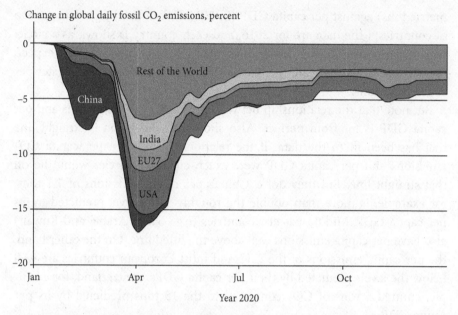

Fig. 5.8 CO_2 Emissions During the COVID-19 Pandemic. The figure shows CO_2 emissions on a monthly basis during 2020, by region.

Source: Le Quéré et al. (2020), United Nations Environment Programme (2020), and Global Carbon Project.

shows CO_2 emissions for different regions of the world and for each month during 2020.

Observe that there was a sharp drop in emissions during March through May of 2020, as countries imposed lock-downs and related measures that prevented people from traveling and even congregating. (In China, the lock-down occurred earlier, first in Wuhan and then in other parts of the country, but the restrictions were mostly relaxed by April, and most of the country returned to normal by June.) Restrictions were still in place from June through the end of the year, but emissions increased, and by September were only about 5 percent lower than the pre-pandemic level in January. And according to the International Energy Agency, by December 2020, global emissions were above the 2019 level.[6]

Can a pandemic save us? Apart from the fact that it would be an extremely unpleasant (and expensive) solution, it would reduce emissions only temporarily. In terms of the long run, it's no help at all.

[6] See https://www.iea.org/articles/global-energy-review-co2-emissions-in-2020.

5.2 CO_2, Methane, and Temperature Change

If global CO_2 emissions begin steadily declining, will the increase in the global mean temperature by the end of the century stay below 2°C? Given all the uncertainties about climate change that we discussed in Chapter 3, we don't know, but we can explore some possibilities. We will do this using a simple model to relate CO_2 emissions to changes in the atmospheric CO_2 concentration, and changes in the CO_2 concentration to changes in temperature. We did this in Chapter 2 using an even simpler model. Now we'll add just a little bit of complexity (and realism) to allow for the lag between changes in the CO_2 concentration and changes in temperature. We will start with the optimistic CO_2 emissions trajectory from Chapter 2, shown in Figure 2.1 on page 29, i.e., starting in 2020 emissions decline steadily, reaching zero by 2100. But we will also consider some other emissions trajectories, some more optimistic and some less so.

In addition to calculating the impact of CO_2 emissions, we will take into account the warming effects of methane emissions. Methane does not contribute as much to climate change as CO_2, but it is important enough that we want to include its impact.

5.2.1 The Warming Effect of CO_2 Emissions

To calculate the impact of any particular path for CO_2 emissions, we need to determine the atmospheric CO_2 concentration that would result from those emissions. To do this, we start with the actual concentration in 1960, and then in each succeeding year add the percentage increase in concentration from emissions that year (after converting from Gt of emissions to ppm of concentration), and subtract the amount that dissipates (at the rate of 0.35 percent per year).[7] Given the path for the atmospheric CO_2 concentration, we determine the impact on temperature from each year's percentage change in concentration. But unlike the calculations in Chapter 2, we account for the time it takes for an increase in the CO_2 concentration to

---

[7] For example, global CO_2 emissions in 1961 were 9 Gt, which added $(9)(0.128) = 1.15$ ppm of CO_2 to the 315 ppm already in the atmosphere. Dissipation in 1961 was $(.0035)(315) = 1.10$ ppm, so the net increase was $1.15 - 1.10 = 0.05$ ppm, making the 1961 concentration $315 + 0.05 = 315.05$ ppm. Emissions in 1962 were 9.4 Gt, which added $(9.4)(0.128) = 1.20$ ppm, and dissipation was $(.0035)(315.05) = 1.10$ ppm, so the 1962 concentration was $315.05 - 1.10 + 1.20 = 315.15$ ppm. The 1963 concentration is calculated in the same way, then the 1964 concentration, and so on.

affect temperature. Estimates of that lag time vary, but a reasonable base case number, which I will use, is 30 years.[8]

Recall that climate sensitivity is the increase in global mean temperature that eventually (around 30 years or more) results from a doubling of the atmospheric CO_2 concentration. As explained in Section 3.4.1 (page 56), there is uncertainty over the true value of climate sensitivity. The "best estimate" according to the IPCC in 2021 is 3.0°C, which is the number that is frequently used when making temperature change projections. I will use this value of 3.0°C, but also examine the implications of lower or higher values for climate sensitivity.

To calculate the change in temperature, we take the percentage increase in the CO_2 concentration each year and multiply by 3.0 to determine its *full long-run impact* on temperature, i.e., the impact it would have after 30 years. But we allow this impact to build up gradually over the 30 years; after one year, the impact is 1/30 of the full impact, after two years it is 2/30 of the full impact, and so on. The details of these calculations are explained in the Appendix to this chapter.

So far we have limited the discussion to the impact on temperature of a rising CO_2 concentration. But we also want to include the impact of methane emissions. I discuss how this is done below, after explaining where methane comes from, and how methane emissions and a changing methane concentration can affect temperature.

5.2.2 Methane Emissions

Figure 5.9 shows global anthropogenic methane emissions along with the atmospheric methane concentration over the past few decades, as measured by the National Oceanic and Atmospheric Administration (NOAA). In order to facilitate the comparison with CO_2 emissions and concentration, methane emissions in this figure are measured in Gigatons per year (the left-hand vertical axis) to facilitate comparison with CO_2 emissions, and the concentration is measured in parts per million (right-hand vertical axis).[9] Keep in mind that anthropogenic (human-generated) methane emissions only account for about 60 percent of *total* methane emissions; methane is also emitted naturally from wetlands, oceans, permafrost, and other sources.

---

[8] See, e.g., Zickfeld and Herrington (2015).

[9] Methane emissions are commonly expressed in terms of *Megatons* (Mt) and the methane concentration is usually expressed in *parts per billion* (ppb). Note that 1 Gt = 1000 Mt, and 1 ppm = 1000 ppb.

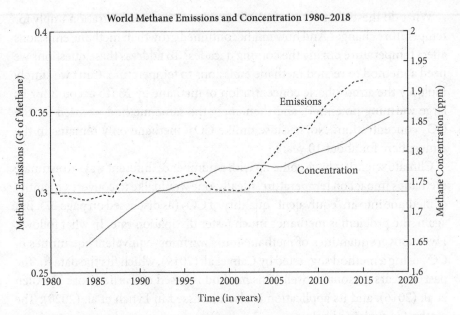

Fig. 5.9 Global Methane Emissions and Atmospheric Concentration. Emissions are in Gigatons (Gt) per year, and concentration is in parts per million (ppm).
Source: National Oceanic and Atmospheric Administration (NOAA).

Recall that global CO_2 emissions have recently been about 37 Gt, which, as Figure 5.9 shows, is about 1000 times the amount of methane emissions. Likewise, the CO_2 concentration has been over 400 ppm, which is more than 200 times the methane concentration. These numbers might suggest that methane is insignificant as a driver of climate change. But one ton of methane has about 28 times the warming potential of a ton of CO_2, and there has been significant growth in methane emissions over the past 20 years. Thus we want to take methane's impact on temperature into account.

As you can see from Figure 5.9, methane emissions were relatively constant from 1980 to 2003, and then increased by more than 20 percent over the next 10 to 15 years. What was the cause of this increase? The most important factor was leakage from increased oil and gas production, and especially the production of oil and gas from shale through the use of hydraulic fracturing (commonly called *fracking*).[10] The atmospheric concentration of methane has increased slowly but steadily over this period, with cumulative growth from 1984 to 2020 of about about 15 percent.

[10]   Alvarez et al. (2018) show that the leakage from fracking alone could explain most or all of the increase in emissions. But Schaefer (2019) attributes much of the increase to other sources, such as agriculture.

What do these changes in methane emissions and concentration imply for temperature change? And how might continued growth in methane emissions affect temperature during the coming decades? To address these questions we need a method to related methane emissions to temperature. Can't we simply multiply the atmospheric concentration of methane by 28 (to account for its large warming potential), and then apply the same method we used for the CO_2 concentration? No, because unlike CO_2, methane only remains in the atmosphere for about 10 years.

Climate scientists have come up with a number of different ways to estimate methane's impact on temperature. Ideally, one would like to convert a quantity of methane into an "equivalent" quantity of CO_2 (as discussed on page 27). But again, the problem is methane's much faster dissipation rate. In what follows, I will convert quantities of methane into "warming-equivalent" quantities of CO_2 using a method suggested by Cain et al. (2019), which fits the data for the past 50 years reasonably well. This method is based on earlier work by Allen et al. (2016), and its application is also discussed in Lynch et al. (2020). The method is explained below.

5.2.3 The Warming Effect of Methane Emissions

First, we need to clarify the statement that a ton of methane has 28 times the warming potential of a ton of CO_2. The number 28 is called the *Global Warming Potential* (GWP) of methane, and it is the answer to the following question: Suppose that today we add one ton of CO_2 and one ton of methane to the atmosphere, and then watch what happens over the next 100 years. The ton of CO_2 and the ton of methane will each cause an increase in temperature, but over the 100 years the temperature increase from the methane will be 28 times as large as the temperature increase from the CO_2.

Note that we could measure the GWP of methane by instead comparing the warming effect of an extra ton of methane and an extra ton of CO_2 over a period of just 50 years, or over a period of 200 years. As a result we sometimes write GWP_H, where H is the time horizon. But the GWP is commonly based on a horizon of 100 years, so we often leave out the H and rather than write GWP_{100}, we just say that the GWP of methane is 28.

Of course by the end of 100 years almost all the methane will have dissipated from the atmosphere, whereas most of the CO_2 will be still be there. Much of the warming effect of a ton of CO_2 occurs gradually over the 100 years, but the warming effect of the ton of methane occurs early on, in the first 10 to 20 years, before the methane has dissipated. In fact, if we used a 20-year horizon to measure the GWP of methane, it would be much higher, about 85. The reason is that over the 20 years, the warming from the ton of CO_2 would have just begun, but the warming from the ton of methane would be close to complete.

Now that we're clear on the meaning of methane's Global Warming Potential, let's turn to the method proposed by Cain et al. (2019) to evaluate the impact of ongoing methane emissions. This method can be applied to any short-lived climate pollutant (SLCP), but we will focus only on methane, which is the most important. The method begins with the GWP and then converts methane emissions over some time period (call it Δt) into a quantity of "warming-equivalent" CO_2 emissions (CO_2-we). It accounts for both the *change* in methane emissions over the time period and the *total amount* of emissions during the time period. Assuming that the GWP horizon is 100 years (i.e., the H in GWP_H is 100) so GWP = 28, it says that the total amount of "warming-equivalent" CO_2 emissions over the time period Δt is given by the following formula:[11]

$$E_{CO_2 we} = 28 \times [r \times \Delta E_M \times 100 + s \times E_M]. \qquad (5.1)$$

Here ΔE_M is the *change* in methane emissions over the time period Δt, and E_M is the *total amount* of methane emissions over that time period. (To get the average annual warming-equivalent CO_2 emissions over the period, just divide $E_{CO_2 we}$ by the time period Δt.) The parameters r and s account for the relative importance of the change in emissions versus the level of emissions. Cain et al. (2019) ran linear regressions to estimate these parameters, and found the best fit was $r = .75$ and $s = .25$.

Let's use this equation to convert global methane emissions over the 35-year period 1980 to 2015 into warming-equivalent CO_2 emissions. Methane emissions in 2015 were about 0.37 Gt, and in 1980 about 0.32 Gt, so the change was $\Delta E_M = 0.05$ Gt and total emissions over this period came to about 12.1 Gt. Then using eqn. (5.1), total warming-equivalent CO_2 emissions over the 35 years were:

$$E_{CO_2 we} = 28 \times [.75 \times 0.05 \times 100 + .25 \times 12.1] = 189.7,$$

---

[11] More generally, for any short-lived climate pollutant, such as nitrous oxide, and any GWP_H, the formula is:

$$E_{CO_2 we} = GWP_H \times \left[r \times \frac{\Delta E_{SLCP}}{\Delta t} \times H + s \times E_{SLCP} \right],$$

where E_{SLCP} is the emissions of the particular SLCP, and GWP_H is the Global Warming Potential for the pollutant over the horizon H.

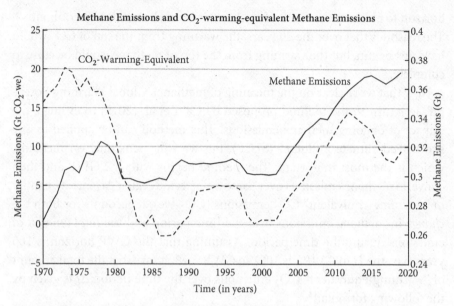

Fig. 5.10 Methane Emissions and CO_2-we Methane Emissions, in Gt per year. Using eqn. (5.1) with overlapping 10-year time intervals, methane emissions (right axis) were converted into CO_2-we emissions (left axis).

which comes to $189.7/35 = 5.4$ Gt per year. For comparison, CO_2 emissions over this period averaged around 27 Gt per year, so in terms of warming-equivalent volumes of GHGs, the contribution of methane was about $5.4/(27 + 5.4) = 5.4/32.4 = 17$ percent. This 17 percent contribution of methane is significant, so we will take it into account when making projections of temperature change.

Figure 5.10 shows historical methane emissions (scale on right) and the corresponding warming-equivalent CO_2 emissions (E_{CO_2we}, scale on left). The E_{CO_2we} emissions were calculated using overlapping 10-year time intervals for Δt in eqn. (5.1). (For example, the 1980 value for E_{CO_2we} is calculated using the data for methane emissions over the ten years 1971 to 1980.) Warming-equivalent CO_2 emissions fluctuate considerably, and can even be negative at times (as during 1986 to 1988); this is because they depend strongly on the *change* in methane emissions, as well as the level.

This leaves one more step: determining how warming-equivalent CO_2 emissions contribute to temperature change. For this last step, I use the fact that there is a rough linear relationship between the temperature change due to CO_2 and cumulative CO_2 emissions to date. This impact on temperature is called the "transient climate response to cumulative carbon emissions" (TCRE), and can be written as TCRE $\approx \Delta T/E_T$, where ΔT is temperature change and E_T is cumulative carbon emissions over the period T. While this relationship was first identified and applied in the context of CO_2 emissions, it can also be applied to warming-equivalent CO_2 emissions coming from methane.[12]

A number of studies have estimated the ratio TCRE, and they are summarized in Knutti, Rugenstein, and Hegerl (2017). Values of the ratio range from about 1.0 to 2.0, with a best estimate of 1.6°C per 1000 Gt carbon. But this number is based on cumulative emissions E_T measured in terms of carbon, not CO_2, so to put this in terms of CO_2 we must divide by 3.66 (the ratio of the mass of CO_2 to the mass of carbon), which yields a value of 0.44°C per 1000 Gt CO_2.[13] Because this approach is based on cumulative emissions, we can determine the impact of methane on temperature by multiplying each year's warming-equivalent CO_2 emissions in Gt by $0.44/1000 = 0.00044$, and then accumulating the resulting temperature changes. (So in 2015, for example, warming-equivalent CO_2 emissions were about 10 Gt, which implies a temperature increase of $10 \times 0.00044 = 0.0044$°C.) That gives us the methane component of temperature change.

It is important to keep in mind that methane can have other detrimental effects, beyond its impact on temperature. For example, through its impact on air quality, it can harm human health and also reduce agricultural and forestry productivity. These impacts are difficult to measure, but they contribute to the benefits of reducing methane emissions.[14]

5.3 Temperature Change Scenarios

In what follows, I will examine several scenarios for future CO_2 emissions and methane emissions, and the implications of those scenarios for temperature change. Climate change involves more than temperature change, but warming is the primary driver of rising sea levels, more frequent and intense hurricanes, and other aspects of climate change. So temperature change is an excellent proxy for climate change more generally.

---

[12] Cline (2020) explains the coincidental linear relationship in detail. Tests and estimates of the TCRE ratio are in Gillett et al. (2013), Matthews et al. (2009), and Matthews et al. (2018).

[13] Fitting a normal distribution to the histogram of about 30 studies conducted over the period 2001 to 2016 yields a mean value of 1.6 and standard deviation 0.39. Dividing the value of 1.6 by 3.66 yields TCRE = 0.44°C per 1000 Gt CO_2, or 0.00044°C per Gt CO_2.

[14] Shindell, Fuglestvedt, and Collins (2017) calculated a social cost of methane (SCM) that purportedly accounts for these effects, along with the warming effects of methane. They arrived at an SCM on the order of $3,000 per metric ton.

The first scenario examined here is the one first presented in Chapter 2, in which CO_2 emissions decline to zero between 2020 and 2100. But now I modify the scenario to include anthropogenic methane emissions, which I will assume also decline to zero between 2020 and 2100. For each scenario, I calculate the warming effects of CO_2 and methane separately, and then add them together to determine the total temperature impact of the two gases. I ignore nitrous oxide and other short-lived GHGs, the impacts of which are generally considered to be minimal. But to the extent that the warming effects of these gases are not minimal, these calculations are conservative, in that they would underestimate the increase in temperature that we can expect.

As explained in Section 5.2.1 and discussed in more detail in the Appendix to this chapter, to calculate the temperature impact of CO_2 emissions, I first determine the atmospheric CO_2 concentration that would result from those emissions. I then determine the impact on temperature from each year's percentage change in concentration, accounting for the roughly 30-year time lag between those changes and the change in temperature. To obtain the temperature impact of methane emissions, I use eqn. (5.1) to convert those emissions into gigatons of warming-equivalent CO_2 emissions, and then multiply by 0.00044.

For the most part I will calculate temperature trajectories using a value of 3.0 for climate sensitivity. Recall that 3.0 is widely used in climate change simulations, and is considered by the IPCC to be the "best estimate" of climate sensitivity. But as discussed in Chapter 3, we don't know the true value of climate sensitivity, so it is important to understand what might happen to temperature change over the coming decades if climate sensitivity turns out to be significantly higher or lower than 3.0. So, we will also calculate temperature trajectories using values of 1.5 and 4.5 for climate sensitivity.

I will consider three different scenarios for global CO_2 emissions and two scenarios for global methane emissions, scenarios that range from the optimistic to the extremely optimistic. The scenarios for CO_2 emissions, which are illustrated in the left-hand panel of Figure 5.11, are as follows:

(1) Starting in 2020, annual emissions of CO_2 fall from 37 Gt to zero by 2100 (as in Figure 2.1 in Chapter 2).
(2) It takes only 40 years to reduce global CO_2 emissions to zero: Starting in 2020, annual emissions of CO_2 fall to zero by 2060, and then remain at zero.
(3) It again takes only 40 years to reduce emissions of CO_2 to zero, but after a delay of 10 years. In this scenario annual emissions remains constant (at 37 Gt) from 2020 to 2030, and then declines to zero by 2070.

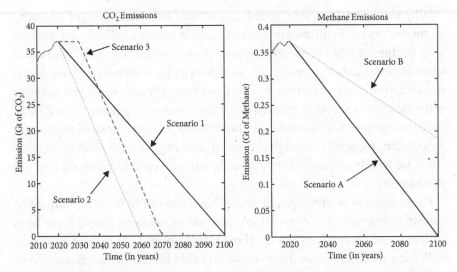

Fig. 5.11 Emission Scenarios. The left panel shows three scenarios for CO_2 emissions: (1) CO_2 emissions fall to zero by 2100. (2) CO_2 emissions fall to zero by 2060. (3) CO_2 emissions stay constant from 2020 to 2030, and then fall to zero by 2070. The right panel shows two scenarios for methane emissions: (A) Methane emissions fall to zero by 2100. (B) Methane emissions decline to half its 2020 level by 2100.

These scenarios for CO_2 emissions might not seem very optimistic for the U.S. and Europe, which have already achieved some emission reductions, and in the case of Europe and the U.K., have pledged to reduce net emissions to zero by 2050. What makes them so optimistic is that they apply to *global* CO_2 emissions. As discussed earlier, CO_2 emissions in much of the world have been rising steadily, and it is unlikely that some of the largest polluters (e.g., India, Indonesia, and China) will reduce their emissions to anywhere close to zero over the next 40 or 50 years. And there is no guarantee that those countries that have pledged to reduce emissions substantially will actually meet those pledges.

Although less important than CO_2, we have seen how methane also contributes to warming. I consider two scenarios for global methane emissions. These scenarios, which are illustrated in the right-hand panel of Figure 5.11, are as follows:

(A) Starting in 2020, annual methane emissions fall from 0.38 Gt (their actual value that year) to zero by 2100.
(B) Starting in 2020, annual emissions fall to half of their 2020 level by 2100, i.e., from 0.38 Gt to 0.19 Gt.

Why would it take so long to reduce methane emissions to zero? And why might we expect methane emissions to fall to only to half of their current level by the end of the century, rather than to zero? The problem is that while some reduction in methane emissions can be achieved relatively easily, reducing those emissions to zero would be extremely difficult. As an example of the easy part, methane emissions increased by over 20 percent from 2003 to 2020, but most of that increase was due to leakage from increased oil and gas production, especially through the use of fracking. That 20 percent increase could be largely eliminated by imposing stricter regulations on oil and gas production.

Other sources of methane emissions, however, would be more difficult to control. For example, a considerable amount of methane comes from farm animals, such as cows and sheep. If the world stopped consuming meat and milk (and wool), much of these emissions could be eliminated. But it is very unlikely that 100 percent, or even 20 percent, of the world's population will become vegan in the coming years.

Also, keep in mind that anthropogenic (human-generated) methane emissions only account for about 60 percent of *total* methane emissions; methane is also emitted naturally from wetlands, oceans, permafrost, and other natural sources. The amount of methane emitted from those sources, however, depends in part on temperature, and will rise as temperatures rise. Since rising temperatures are due to human activity, the resulting increase in methane emissions can be considered anthropogenic.

What matters in the end is how methane emissions can affect climate, and in particular temperature. Because methane dissipates rapidly from the atmosphere (most of the methane emitted today will be gone in 10 years), if methane emissions stayed constant at today's level, the impact on temperature would be quite limited. As explained in Section 5.2.3, it is the change in methane emissions that matters most, as opposed to the level of emissions. (Recall from equation (5.1) that the quantity of "warming-equivalent" CO_2 emissions, CO_2-we, depends largely on the change in methane emissions as opposed to the level of emissions.)

5.3.1 Changes in Temperature

What changes in temperature can we expect for these different scenarios? The corresponding temperature trajectories, calculated out to the year 2100, are shown in Figure 5.12. The left-hand panel of the figure is based on the first (and more optimistic) scenario for methane, i.e., annual emissions fall from

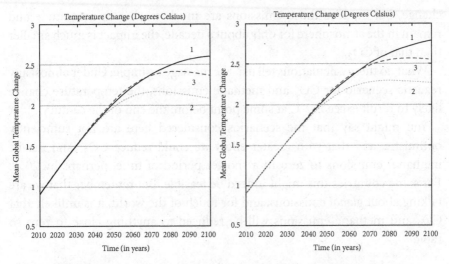

Fig. 5.12 Resulting Temperature Trajectories. Both panels show the temperature trajectories that result from each of the three scenarios for CO_2 emissions shown in the left panel of Figure 5.11: (1) CO_2 emissions fall to zero by 2100; (2) emissions fall to zero by 2060; (3) emissions are constant from 2020 to 2030 and then fall to zero by 2070. In the left panel, we assume methane emissions go to zero by 2100 (Scenario A in the right panel of Figure 5.11); in the right panel we assume methane emissions decline to half its 2020 level by 2100 (Scenario B).

0.38 Gt to zero by 2100. The trajectories labeled 1, 2, and 3 correspond to the three scenarios for annual CO_2 emissions: (1) CO_2 emissions fall to zero by 2100. (2) CO_2 emissions fall to zero by 2060. (3) CO_2 emissions stay constant from 2020 to 2030, and then fall to zero by 2070.

As the left-hand panel of Figure 5.12 shows, for all three scenarios the temperature change exceeds 2°C at some point before the end of the century. Scenario (2) is the most optimistic of the three; it has CO_2 emissions reaching zero within the next 40 years, and as a result, the temperature increase only reaches about 2.2°C. But if the decline in global CO_2 emissions to zero only begins in 2030 (and then takes 40 years), the temperature increase exceeds 2.4°C. And if it takes until the end of the century for CO_2 emissions to reach zero, the temperature increase exceeds 2.5°C.

What if the decline in methane emissions is more moderate, i.e., emissions only decline to half the 2020 level by 2100? As the right-hand panel of Figure 5.12 shows, the temperature increases are larger, but the effect of higher methane emissions is quite limited. For all three CO_2 scenarios, the maximum increase in temperature is about 0.2 to 0.3°C larger than on the left-hand panel. As explained earlier, methane emissions contribute to climate change, but because those emissions are much smaller in magnitude and remain in the atmosphere for only about a decade, the impact is much smaller than that of CO_2.

What do these calculations tell us? The message is simple: Under almost any realistic scenario for CO_2 and methane emissions, the temperature change likely to result exceeds 2°C at some point before the end of the century.

You might say that the scenarios considered here are not sufficiently optimistic, and that with greater effort we could reduce CO_2 (and maybe methane) emissions to zero in a shorter period of time, perhaps by 2050. For some countries this might indeed be feasible. But remember that we are talking about *global* emissions, and for much of the world, it is unlikely that CO_2 and methane emissions will be reduced to anything close to zero so rapidly.

5.3.2 Implications of Uncertainty

The temperature trajectories shown in Figure 5.12 are based on a middle-of-the-road estimate of 3.0 for climate sensitivity. But this estimate of 3.0, while widely used in simulations and temperature projections, is just an estimate, and we know that the actual value of climate sensitivity is uncertain. How sensitive are these results to uncertainty over the value of climate sensitivity? To address this question, we repeat the two most optimistic scenarios (numbers 2 and 3 in Figure 5.12), but this time using three different values of climate sensitivity: 1.5, 3.0, and 4.5.[15]

There is also uncertainty over the time lag between changes in the atmospheric CO_2 concentration and the change in temperature; while 30 years is a middle-of-the-road estimate for that lag, it could be as low as 20 or as high as 50, and depends in part on both the level and the change in the CO_2 concentration. Likewise there is uncertainty over the annual dissipation rate for atmospheric CO_2; although .0035 best fits the data, the actual rate could be as low as .0025 or as high as .0050. However, for simplicity I will ignore these additional uncertainties and focus only on uncertainty over climate sensitivity. For the results that follow, the time lag is fixed at 30 years and the annual dissipation rate is .0035.

The results are shown in Figure 5.13. Both panels show temperature trajectories for three value of climate sensitivity, $S = 1.5$, 3.0, and 4.5. The left

---

[15] These numbers span the IPCC's "most likely" range of values for climate sensitivity until their most recent report; in 2021 they narrowed the "most likely" range to 2.5 to 4.0.

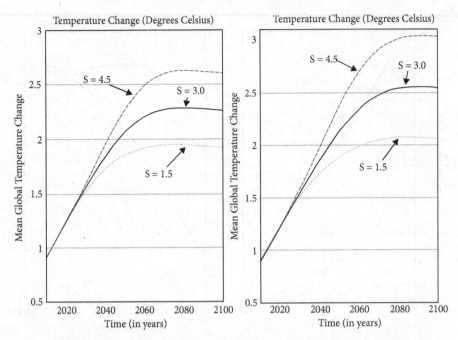

Fig. 5.13 Temperature Trajectories for Alternative Values of Climate Sensitivity. Both panels show temperature trajectories for three value of climate sensitivity, $S = 1.5$, 3.0, and 4.5. In the left panel we assume that CO_2 emissions fall to zero by 2060. In that case, if $S = 1.5$ the temperature increase stays just below 2°C, but if $S = 4.5$, the increase exceeds 2.6°C. In the right panel emissions are assumed to stay constant from 2020 to 2030 and then fall to zero by 2070. This scenario is a bit less optimistic, but now the temperature increase exceeds 2°C even if $S = 1.5$, and exceeds 3°C if $S = 4.5$.

panel applies to the most optimistic scenario shown in Figure 5.12, i.e., CO_2 emissions fall to zero by 2060. In that case, if $S = 1.5$ the temperature increase stays just below 2°C, but if $S = 4.5$, the increase exceeds 2.6°C. In the right panel emissions are assumed to stay constant from 2020 to 2030 and then fall to zero by 2070. This scenario for CO_2 emissions is just a little less optimistic, but now the temperature increase exceeds 2°C even if $S = 1.5$, and exceeds 3°C if $S = 4.5$. (In both panels we assume that methane emissions declines to half its 2020 level by 2100.)

These results show—as expected—that whatever the trajectory for CO_2 emissions over the next several decades, temperature change depends critically on the actual value of climate sensitivity. At this point we don't know the actual value, and there is a wide range of plausible numbers. If we are lucky and it turns out that climate sensitivity is at the low end of the range, *and* we are able to quickly and sharply reduce global CO_2 emissions, we might indeed be able to keep the temperature increase below 2°C. But if we are not

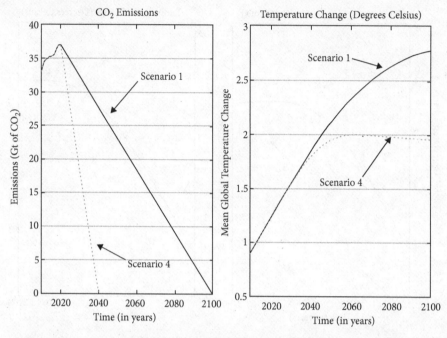

Fig. 5.14 The 2°C Scenario. The left panel shows two scenarios for CO_2 emissions. The first, which replicates Scenario 1 from Figure 5.12 (and is labeled Scenario 1), has emissions fall to zero by 2100. The second (labeled Scenario 4) has emissions fall to zero in just 20 years, i.e., by 2040. The resulting temperature trajectories are shown in the right panel.

so lucky and climate sensitivity is at the middle or high end of the range, the temperature increase will almost surely exceed 2°C, even if global CO_2 emissions start declining immediately and reach zero by 2060.

Suppose the correct value for climate sensitivity is the widely used one of 3.0. Letting our optimism run wild, is there any scenario for global CO_2 emissions that would keep the temperature increase below 2°C? Yes. If emissions fell to zero in just 20 years, i.e., by 2040, and then stayed at zero, the temperature increase would be just a hair below 2°C. This is illustrated in Figure 5.14. The left panel of the figure shows two scenarios for CO_2 emissions. The first, shown for comparison because it replicates Scenario 1 from Figure 5.12 (and is labeled Scenario 1), has emissions fall to zero by 2100. The second (labeled Scenario 4) has emissions fall to zero in just 20 years, i.e., by 2040. The resulting temperature trajectories are shown in the right panel.

Scenario 4 represents unbounded enthusiasm, because it is inconceivable that *global* CO_2 emissions could be reduced to zero so fast (and then remain at zero). However, it does result in a temperature increase that stays just below

2°C. But remember that this calculation assumes climate sensitivity is 3.0 or lower; if climate sensitivity were a little higher, say, 3.5, the temperature increase would exceed 2°C.

All of the scenarios we've looked at involve reducing CO_2 emissions as the way to limit temperature increases. But have we tied our hands, so to speak, by ignoring an important alternative to reducing emissions—carbon removal and sequestration? Recall that the idea is to remove CO_2 from the atmosphere ("removal") and then store it in some permanent way ("sequestration").

How can we remove CO_2 from the atmosphere? Trees absorb CO_2, so one option is to plant trees. Currently we are cutting down trees, not planting them, but with the right incentives reforestation, and planting new forests, is certainly possible. I will discuss forestation in more detail later, but the basic problem is that it would take a huge number of new trees to absorb enough CO_2 to make a significant difference in net emissions. What about using new and evolving technologies to extract CO_2 from power plant emissions or directly from the atmosphere and storing it underground? I will also discuss this option in more detail later, but the problem is that we simply don't have the technology to do carbon removal and sequestation on a large scale, at least not at anything approaching a reasonable cost.

So where does this leave us? We should push hard to reduce CO_2 emissions as much as possible, using whatever policy instruments are available (and sensible). And we should do this as part of an international agreement that commits other countries to sharply reduce emissions as well, because it is *global* emissions that matter. But at the same time we have be aware that despite our best intentions, it is very possible—in fact likely—that the global mean temperature will rise well above 2°C. We don't know what the impact of such a temperature increase will be, but the impact might be severe. We need to plan and take action accordingly.

5.4 Rising Sea Levels

So far we have spoken only about temperature change, which is indeed the fundamental problem with an increasing atmospheric CO_2 concentration. But one of the major concerns about warming is that it could lead to rising sea levels and wide-spread flooding. Why would sea levels rise? Because higher temperatures can cause sea water to expand, and can cause glaciers to melt and fragment. Let's look at what might lie ahead for sea levels.

There is evidence that sea levels have already risen somewhat, but the concern is that increases in the global mean temperature could them to rise much more. By how much should we expect sea levels to rise over the coming decades? That depends on the how much temperatures rise. OK, let's assume that the global mean temperature increases by 3°C by the end of the century. What, then, would happen to sea levels? The answer is . . . , well, by now you can probably complete the sentence. The answer is: We don't know. We don't know how much sea levels will rise if the global mean temperature rises by 2°C, by 3°C, or any other amount of warming.

Why don't we know how much sea levels will rise if the temperature rises by, say, 2°C? For the same reason that we don't know what any particular increase in the atmospheric CO_2 level will do to the temperature. The physical systems that relate CO_2 concentration to temperature and temperature to sea levels are just too complex and poorly understood. What can we say about sea levels? A number of studies have suggested possible ranges for sea level increases, just as we have ranges for the value of climate sensitivity.

Figure 5.15 shows five projections of sea level rise by 2100 for different temperature increases. Two of the projections were part of the IPCC's Fourth (2007) and Fifth (2013) Assessments (Solomon et al. (2007) and Stocker et al. (2013) respectively), and the other three are from Vermeer and Rahmstorf (2009), Kopp et al. (2014), and Mengel et al. (2016). Each projection is accompanied by an error bound. (For example, Vermeer and Rahmstorf (2009) estimate that a 2°C increase in global mean temperature would result in a 0.8 to 1.3 meter sea level rise, with a best estimate of 1.0 meters.)

What does Figure 5.15 tell us? First, that the projections are very different, even those made at about the same time. For example, IPCC (2007) estimates the sea level rise to be between 0.2 and 0.5 meters, even for temperature increases as high as 4°C, while the estimates by Vermeer and Rahmstorf (2009) are centered around 1.0 to 1.5 meters. Second, the projected sea level rise doesn't depend much on the size of the temperature increase, which at the very least is counter-intuitive. For example, IPCC (2013) projects a sea level rise of 0.44 meters for a temperature change of 1°C, 0.55 meters for a 2.2°C change, and 0.74 meters for a 3.7°C change.

So what will happen to sea levels? The studies summarized in Figure 5.15 don't give us much guidance. The projections vary widely and don't tell a clear story. Probably the most we can infer from Figure 5.15 is that even if warming is substantial (greater than 2 or 3°C), the sea level rise might be less than half a meter, or as much as 1.4 meters. Furthermore, the studies summarized here

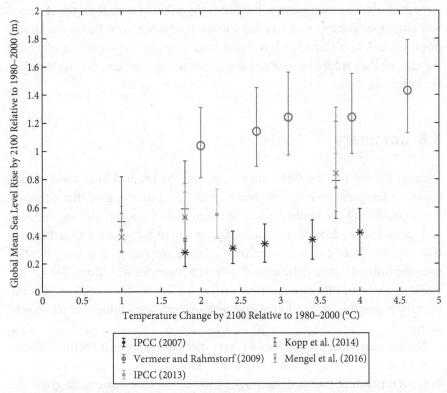

Fig. 5.15 Projections of Sea Level Increase. Figure shows estimated end-of-century global mean sea level rise and associated error bounds as a function of end-of-century temperature increase. IPCC 2007 refers to Solomon et al. (2007) and IPCC 2013 refers to Stocker et al. (2013), respectively. The other studies are Vermeer and Rahmstorf (2009), Kopp et al. (2014), and Mengel et al. (2016).

are projections of *global* sea level rise, and local changes in sea levels can be very different from the global average, and even less predictable.[16]

What should we make of the fact that there is so much uncertainty over future sea levels? Some people might say that we should wait and see what happens, rather than taking costly actions now, such as building sea walls or dikes to reduce the possible impact of rising sea levels. But of course that's the same argument as holding off on a carbon tax or other measures to reduce CO_2 emissions until we learn more about climate sensitivity and about the

[16] See, for example, Kopp et al. (2014) and Stammer et al. (2013). Hansen et al. (2016) paint a more pessimistic picture, and argue that even with a temperature increase limited to 2°C, there is a good chance that ice sheets will disintegrate, raising sea levels by several meters.

impact of higher temperatures. It ignores the value of insuring against a very bad outcome. The fact that you don't know if or when your house may flood doesn't mean you shouldn't buy flood insurance. Sea levels might rise only slightly, or they might rise a great deal, and acting now can protect us from the latter outcome.

5.5 Summary

What is the outlook for GHG emissions, and the implied likelihood of preventing a temperature increase greater than 2°C? I have argued that although some countries have made pledges (or even passed laws) requiring net-zero CO_2 emissions by 2050, it is unlikely that we will reduce *global* CO_2 emissions that rapidly. Note the word "unlikely." There are substantial uncertainties over the future climate policies that different countries will adopt. But for the world as a whole, although reaching zero emissions well before the end of the century is possible, it is unlikely, and certainly not something that we should count on.

The alternative scenarios for CO_2 and methane emissions (some realistic, some less so) that we examined paint a picture for changes in the global mean temperature that is discouraging, to say the least. Of course given the uncertainty over climate sensitivity (and other aspects of the climate system), it is possible that things will work out better than expected. For example, if the true value of climate sensitivity turns out to be much smaller than the widely used estimate of 3.0, the temperature increase might indeed stay below 2°C. But if the true value of climate sensitivity turns out to be larger than 3.0, the temperature increase could end up well above 2°C, even if the entire world adopts and implements aggressive emission reduction policies. And with less aggressive emission reduction policies, it is very possible that the temperature increase will exceed 3°C.

Some readers might think that the picture I've painted is too pessimistic, or even defeatist. Perhaps it is. As I said, it is hard to predict the climate policies that various countries will adopt over the coming decades, and we might be pleasantly surprised by what countries ranging from the U.S. to China and India end up doing. And we might also be pleasantly surprised to eventually learn that the true value of climate sensitivity is lower than expected. At issue is whether we should count on being pleasantly surprised. In the next chapter I explain why doing so is not just naive, but dangerous.

5.6 Further Readings

This chapter looked in more detail at the challenges and prospects for substantially reducing GHG emissions, and the implications for temperature change through the end of the century. I have argued that the possibility (if not likelihood) of a large temperature increase means that we need to prepare accordingly and focus soon in various forms of adaptation. But that does not mean that we should give up on a net zero emission target, and numerous studies have addressed the steps that can be taken to reach this target.

- A consortium of researchers, the "Sustainable Development Solutions Network," have assembled a very detailed report that describes the actions that can be taken to sharply reduce GHG emissions. See SDSN 2020 (2020).
- Another consortium of researchers, many based at Princeton University, have likewise assembled a report describing in detail actions to reduce GHG emissions. See Larson et al. (2020).
- For a focus on the United States, Heal (2017b) provides a detailed analysis of how GHG emissions could be by reduced 80 percent by 2050, and why it would be difficult to reduce emissions further than this.
- What is the argument for a 2°C limit on warming? Actually, some have argued that the correct limit should be only 1.5°C. How much worse would 2°C of warming be compared to 1.5°C? For a detailed analysis, see the report from Intergovernmental Panel on Climate Change (2018).
- The M.I.T. Joint Program on Global Change has used its model of the climate system to simulate CO_2 emission trajectories that could prevent the temperature increase from exceeding 2°C, and has also examined the implications of alternative CO_2 emission trajectories. The results and methodology are described in Sokolov et al. (2017).
- The scenarios for CO_2 and methane emissions considered in this chapter might be viewed as overly optimistic because they ignore the possibility of a melting permafrost. See Schuur et al. (2015) and Knoblauch et al. (2018) to learn more about this problem.
- Finally, for more optimistic views regarding the amount of warming we might expect and the impact that warming might have, see Lomborg (2020) and High-Level Commission on Carbon Prices (2017). For a more pessimistic view, see Stern (2015).

5.7 Appendix to Chapter 5: Temperature Scenarios

In Section 5.3 we presented temperature trajectories out to the year 2100 based on alternative scenarios for CO_2 emissions and methane emissions, and briefly described how those emissions are translated into temperature changes. This Appendix provides more detail on how these calculations are done, first for CO_2 and then for methane.

CO_2 Emissions

How does an increase in the atmospheric CO_2 concentration affect global mean temperature? Unlike the calculations in Chapter 2, here we account for the time it takes for an increase in the CO_2 concentration to affect temperature. Estimates of that lag time vary, but a reasonable base case number, which I use here, is 30 years.

Of course we also need a value for climate sensitivity, i.e., the increase in temperature that eventually results from a doubling of the atmospheric CO_2 concentration. As explained in Section 3.4.1, there is considerable uncertainty over the true value of climate sensitivity. I use 3.0, which is in the middle of

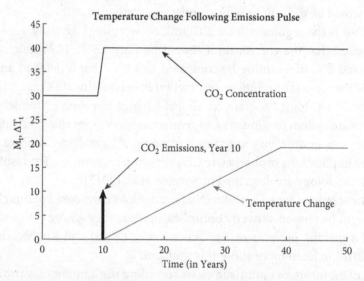

Fig. 5.16 Illustrative Example of Effect of CO_2 Emissions in Year 10 on Temperature. In this hypothetical example, a large quantity (a "pulse") of CO_2 is emitted in Year 10, which immediately increases the CO_2 concentration that year. However, the full impact on temperature takes 30 years.

the "most likely" range according to the IPCC, but we will also examine the temperature change implications of other values for climate sensitivity.

To calculate the change in temperature, we take the percentage increase in the CO_2 concentration each year and multiply by the value of climate sensitivity, say 3.0, to determine its *full long-run impact* on temperature, i.e., the impact it would have after 30 years. But we allow this impact to build up gradually over the 30 years; after one year, the impact is 1/30 of the full impact, after two years it is 2/30 of the full impact, and so on.

This is illustrated in Figure 5.16, which shows hypothetical CO_2 emissions, the CO_2 concentration, and the change in temperature over time. Here we assume that the *only* CO_2 emissions occur as a large "pulse" in Year 10, so that the CO_2 concentration is constant until year 10, then jumps by the amount of that year's emissions, and remains constant at the new higher level. (We are ignoring dissipation in the figure.) What happens to temperature? Initially nothing, because it takes time for the increase in CO_2 concentration to affect temperature (via climate sensitivity). How much time? Recall that it can take over a century for the climate system to reach a new equilibrium after an increase in the CO_2 concentration, but most of the effect on temperature occurs within a 20- to 40-year window. Here we have assumed that it takes 30 years for the full effect to occur, with temperature increasing linearly over the 30 years. Thus the temperature increases from Year 10 to Year 40, and then remains constant from Year 40 onwards.

In what follows, we denote emissions of CO_2 at time t by E_t, the atmospheric concentration of CO_2 by M_t, and the dissipation rate of atmospheric CO_2 by δ. Finally, let S be the value of climate sensitivity. Then given E_t and M_t, the concentration of CO_2 in the following year $t + 1$ is just:

$$M_{t+1} = (1 - \delta)M_t + E_{t+1} \tag{5.2}$$

We set the dissipation rate to $\delta = .0035$, which provides a good fit to the historical data.

We want the change in temperature, ΔT_t, starting in year $t = 1$. To get this, I assume that prior to the start date (say 1950 or earlier) there was very little change in M_t, so that any temperature impacts can be ignored. Thus $\Delta T_0 = 0$. Denoting the concentration at the start date by M_0, in year 1,

$$\Delta T_1 = \frac{M_1 - M_0}{M_0} \cdot \frac{S}{30},$$

In the following year, ΔT_2 has two components: the ongoing (and now larger) impact of the increased concentration due to E_1, and the additional impact of the increase in concentration due to E_2

$$\Delta T_2 = \frac{M_1 - M_0}{M_0} \cdot \frac{2S}{30} + \frac{M_2 - M_1}{M_1} \cdot \frac{S}{30}$$

Likewise

$$\Delta T_3 = \frac{M_1 - M_0}{M_0} \cdot \frac{3S}{30} + \frac{M_2 - M_1}{M_1} \cdot \frac{2S}{30} + \frac{M_3 - M_2}{M_2} \cdot \frac{S}{30}$$

And for $k \leq 30$,

$$\Delta T_k = \left[\frac{M_1 - M_0}{M_0} \cdot \frac{k}{30} + \frac{M_2 - M_1}{M_1} \cdot \frac{(k-1)}{30} + \dots + \frac{M_{k-1} - M_{k-2}}{M_{k-2}} \cdot \frac{2}{30} \right.$$
$$\left. + \frac{M_k - M_{k-1}}{M_{k-1}} \cdot \frac{1}{30} \right] S$$

For $k > 30$, the temperature change will include the full impact of emissions prior to 30 years earlier, and the partial impacts of more recent emissions:

$$\Delta T_k = \left[\frac{M_1 - M_0}{M_0} + \frac{M_2 - M_1}{M_1} + \dots + \frac{M_{k-30} - M_{k-31}}{M_{k-31}} + \frac{M_{k-29} - M_{k-30}}{M_{k-30}} \right] S$$
$$+ \left[\frac{M_{k-28} - M_{k-29}}{M_{k-29}} \cdot \frac{29}{30} + \frac{M_{k-27} - M_{k-28}}{M_{k-28}} \cdot \frac{28}{30} + \dots + \frac{M_k - M_{k-1}}{M_{k-1}} \cdot \frac{1}{30} \right] S$$

To summarize, we start with an estimate of the initial atmospheric CO_2 concentration, values for historical CO_2 emissions (through 2020), and projections of CO_2 emissions for some scenario. We use the equations above to calculate the atmospheric CO_2 concentration and the temperature change (relative to the base year $t = 0$) in each year.

Methane

Working with overlapping 10-year time intervals, we first use eqn. (5.1) to convert ΔE_M (the change in methane emissions over the 10 years) and E_M (total methane emissions over that period) to get the total warming-equivalent CO_2 emissions over the 10 years, E_{CO_2we}. Dividing by 10 gives the annual average warming-equivalent CO_2 emissions from methane.

The TCRE method is used to determine the impact of E_{CO_2we} on temperature. Over a 10-year interval, TCRE $\approx \Delta T / E_{CO_2we}$, where ΔT is the temperature change over the interval and E_{CO_2we} is the corresponding warming-equivalent CO_2 emissions. Knutti, Rugenstein, and Hegerl (2017) has surveyed studies that have estimated the ratio TCRE, and they are summarized as a histogram in Figure 5.17. The histogram gives a rough fit to a normal distribution, with a mean of 1.6°C per 1000 Gt carbon. To put this in terms of CO_2 rather than carbon we divide by 3.66 (the ratio of the mass of CO_2 to the mass of carbon), which yields a value of 0.44°C per 1000 Gt CO_2, or 0.00044°C per Gt CO_2. Thus the impact of methane on temperature is found by multiplying each year's warming-equivalent CO_2 emissions in gT by 0.00044, and then accumulating the resulting temperature changes.

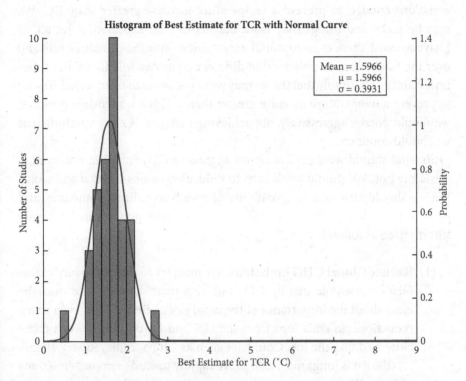

Fig. 5.17 Histogram of studies showing estimates of TCRE for methane. Estimates of the ratio are from Knutti, Rugenstein, and Hegerl (2017), and a normal distribution was fitted to the histogram. The mean of the distribution is 1.60.

6
What to Do: Reducing Net Emissions

I have painted a rather pessimistic picture of climate change. I argued that it is unlikely—not impossible, but unlikely—that we will reduce *global* CO_2 emissions enough to prevent a temperature increase greater than 2°C. We may be lucky and things may work out better than expected. After all, as I have stressed, there are substantial uncertainties over the climate system and over the future climate policies that different countries will adopt. But those uncertainties also imply that things may work out *worse* than expected. Trying to prevent a temperature increase greater than 2°C is a worthwhile goal that we should pursue aggressively, but achieving that goal is not something that we should count on.

So what should we do? Give up on aggressive CO_2 emission reductions? Certainly not. We should work hard to reduce emissions—*global* emissions. But we should also face the possibility of an adverse climate outcome, and prepare accordingly. The main elements of climate policy can be briefly summarized as follows:

(1) **Reduce Global GHG Emissions.** We must try to sharply reduce global GHG emissions; mainly CO_2 but also methane. And we must be clear about the importance of the word *global*. Reductions, even sharp reductions, in emissions from the U.S. and Europe alone won't come close to doing the job. Countries such as China, India, Russia, Brazil, ... (the list is long) must also move rapidly towards zero net emissions of CO_2. This means that reducing emissions must be part of an international agreement, and one that can realistically be enforced.

(2) **Reduce Emissions as Efficiently As Possible.** We should adopt climate policies that are *efficient*, which means that GHG emissions are reduced at the lowest possible cost. Study after study after study has shown that the most efficient and straightforward way to achieve this goal is through the use of a carbon tax. But to the extent that the adoption of a sufficiently large carbon tax is politically infeasible, we should also pursue other options such as directed subsidies

Climate Future: Averting and Adapting to Climate Change. Robert S. Pindyck, Oxford University Press.
© Oxford University Press 2022. DOI: 10.1093/oso/9780197647349.003.0006

and government mandates. And despite the objections of some environmentalists, we have to consider the expanded use of nuclear power to generate electricity.

(3) **Pursue Options to Remove Carbon from the Atmosphere.** To the extent possible, we can try to remove CO_2 from the atmosphere, and capture and store CO_2 emissions from power plants, thereby reducing net emissions. This will be difficult, and barring some major new innovations, is unlikely to contribute that much to solving our climate problem. Planting trees is an option, but we will see that it would take huge number of trees to make much of an impact. Given the currently available technologies, we shouldn't expect carbon removal to make a major dent in the growing atmospheric CO_2 concentration. But even a minor dent is better than nothing, and in the future a major dent might be possible if we invest now in the R&D to develop new technologies for carbon removal and storage.

(4) **Invest in Adaptation.** We must acknowledge the fact that despite our best efforts, emissions will not fall fast enough and the atmospheric CO_2 concentration will continue to rise. This could mean an increase in the global mean temperature that is greater, perhaps much greater, than the widely cited 2°C limit. It likewise means that we may face rising sea levels, more frequent and stronger hurricanes and storms, and other adverse climate effects. We must prepare for the possibility of that outcome. How to prepare? By investing now in *adaptation*. Adaptation includes everything from the development of new heat-resistant crops to the construction of sea walls to the use of solar geoengineering. (Yes, geoengineering, the very thought of which drives some environmentalists crazy, but read on before coming to any conclusions.)

In this chapter I will discuss ways to reduce CO_2 emissions. I will explain why directed subsidies and government mandates can be quite effective, but a carbon tax—especially one that is part of an international agreement—is the most efficient policy tool at our disposal. I will also discuss the use of nuclear power to generate electricity. Then I will turn to the prospects for carbon removal and sequestration (i.e., storage), and examine two widely promoted approaches—planting trees, and removing CO_2 directly from the air and from the smokestacks of coal-burning power plants.

In the next chapter I will discuss the other critical leg of climate policy, namely adaptation. I will explain how agriculture has already adapted to climate change. I will then focus on ways to counter rising sea levels and more intense hurricanes, and ways to reduce the warming effects of a rising atmospheric CO_2 level, i.e., the use of geoengineering.

6.1 How to Reduce Emissions

Sharply reducing emissions is clearly Number 1 on the climate to-do list. But how? Reducing CO_2 emissions boils down to reducing carbon intensity, i.e., the amount of CO_2 that results from each dollar of GDP. (We could also reduce emissions by reducing GDP, e.g., by engineering a major recession, but that's not a happy alternative.) As explained in Chapter 3, a decline in carbon intensity can result from a decline in energy intensity (the amount of energy used to produce each dollar of GDP), and/or an improvement in energy efficiency (the amount of CO_2 emitted from the use of a unit of energy).

Energy intensity and energy efficiency can both be affected by government policy, and in a sense that is what much of climate policy is all about. Consider a carbon tax, which would raise the price of burning carbon. Since most of the energy we use comes from burning carbon, the tax would reduce our use of energy by making it more expensive, and thus would cause a decline in energy intensity. And of course other policy options, such as automobile fuel efficiency standards, and "green" building and appliance codes, could also be used to reduce energy intensity.

A carbon tax would also create incentives to produce the energy we use with less carbon, and thereby improve energy efficiency. For example, a BTU of energy obtained from burning natural gas produces about half as much CO_2 compared to a BTU obtained by burning coal, so a carbon tax would make coal relatively more expensive then natural gas, and thereby result in a more rapid shift away from coal. (The shift away from coal could also be achieved via direct regulations over the construction of new power plants.) Likewise, a BTU of energy obtained from wind power burns no carbon, and thus becomes more economical once a carbon tax is in place.

Policies to reduce carbon intensity seem simple enough, so what are the impediments to their rapid adoption? One important concern is what a carbon tax, fuel efficiency standards, and other policy measures would cost in terms of reduced consumption (private and public) and consumption growth. Those costs are unclear. We know that marginal abatement costs rise as the amount of abatement increases—the cost of a 20 percent emission reduction is more than twice the cost of a 10 percent reduction—but we don't know what those costs actually are. Estimates of current CO_2 abatement costs vary widely, and *future* abatement costs are even more uncertain because we can't predict the cost reductions that would result from technological change.

And even if the cost of strong emissions abatement is moderate, are the policies that we would need politically feasible? How would voters respond to the prospect of a carbon tax that would raise their costs of driving, home heating, etc.? And how would they respond to the impacts major subsidies would have on government budgets? The answers will differ from country to country (and in many countries the views of voters are irrelevant), complicating the adoption of policies that are part of an international agreement. But an international agreement is essential, because as I've said repeatedly, it is global emissions that matter, not the emissions of the U.S. and Europe.

The need for an international agreement brings us to the free-rider problem—let other countries reduce emissions, and we'll still benefit— which reduces the political feasibility of strong abatement policies in many countries. One purpose of an international agreement is to overcome the free-rider problem, but how can we actually get such an agreement, and make it stick? As discussed below, besides its other benefits, a carbon tax— as opposed to subsidies or government mandates—can help overcome the free-rider problem and achieve a binding international agreement.

6.1.1 A Carbon Price

Climate policy in the U.S., Europe, and many other countries revolves around subsidies and government mandates. Examples are subsidies (e.g., via tax credits) for the purchase of electric cars, and average fuel efficiency standards to reduce gasoline consumption by non-electric cars. A related policy proposal (in the U.S.) would have the government pay for a large number of charging stations, which would make electric cars more attractive. Those are examples, and there are many others. What's missing is the simplest and most efficient policy tool: pricing carbon to reflect its true cost. One way to do this is by imposing a carbon tax. Another way is to use a cap-and-trade system, which I'll say more about later, in Section 6.1.4. A carbon tax is straightforward and has other advantages, so I'll begin with that. Things may change, but so far (at least in the U.S.) a carbon tax is rarely discussed as a key part of climate policy.

Why are economists so fixated on the use of a carbon tax to reduce CO_2 emissions? Why not rely (or rely more) on subsidies and direct regulations?

And why is the public so opposed to a carbon tax? What's wrong with pricing carbon in a way that reflects its true cost?

In simple terms, economists' preference for a tax is based on the notion of asking people to pay for things they use or consume. Most people planning to get a new car would not be surprised to learn they have to pay for it. And most people expecting to take their new car on a long trip would not be surprised to learn they have to pay for the gasoline it uses.

We also ask people to pay for any damage they inflict on other people. Suppose you weren't paying attention and drove your new car into your neighbor's car, which happened to be parked on the street. You would expect to have to pay for the resulting damage, either directly, or more commonly, via your insurance policy. But what about the damage caused by the exhaust from your car? That exhaust contains (among other things) CO_2, which contributes to harmful climate change. Shouldn't you have to pay for that damage as well? An economist would say yes, of course you should.

That's the basic idea. If your consumption of a gallon of gasoline causes damage to other people—in this case society at large—you should pay for that damage. And how would you pay for it? By way of a tax on gasoline, a tax just sufficient to cover the damage you have caused society by burning that gallon of gasoline.

A carbon tax would pay for the damage caused by the consumption of gasoline, but also the damage caused by burning carbon and emitting CO_2 in any other way. Recall from Chapter 3 that the cost to society of emitting one additional ton of CO_2 is referred to as the Social Cost of Carbon (SCC). It is a "social cost," i.e., an externality, because the households or firms that emit the CO_2 don't bear this cost; instead society does. The SCC is the basis for a carbon tax. Imposing a tax based on the SCC would correct for the fact that households and firms don't bear the full cost of their CO_2 emissions. If you emit a ton of CO_2 and that results in a cost to society of $100, then you should be asked to pay that cost. A $100 per ton carbon tax would correct the problem—you would have to pay for the damage your ton of CO_2 emissions has caused.

6.1.2 Government Subsidies

OK, there is a $100 social cost to burning a ton of carbon, so we want to reduce the total amount of carbon that we burn. Yes, a carbon tax will do the job, but couldn't we instead use a government subsidy to reduce CO_2 emissions? We could subsidize solar panels, or windmills, or electric cars, or whatever is most popular. Wouldn't that boil down to the same thing, and we'd avoid having to introduce the "t" word that politicians (and much of the public) abhor.

Yes, subsidies of that sort could indeed be used to reduce the amount of carbon we burn, and they already are being used. But the cost to society of that reduction would be higher than if it were done using a carbon tax. To see why, suppose the alternative to the carbon tax is a subsidy for electric cars. To the extent that electricity is generated using renewables as opposed to fossil fuels, electric cars would emit less CO_2 than gas-powered cars.

Who benefits from this subsidy? First, companies that produce electric cars. The subsidy reduces their production costs, and will increase sales because they will be able to charge a lower price for the cars. The result is a gain for consumers as well as a gain (in the form of higher profits) for the electric car companies. What share of the total gain do consumers get? The answer depends on the relative price elasticities of demand and supply for electric cars. If supply is relatively inelastic, which is likely the case for electric cars because of limited production capacity, most of the gain will go to the companies.[1] Still, consumers do benefit, but now ask who are the consumers that will benefit the most? They are the ones who are most likely to buy electric cars. This may change, but so far they have been overwhelmingly people with high incomes.

By how much would CO_2 emissions fall in response to an electric car subsidy? That depends on how much of a shift from gasoline-powered to electric cars the subsidy causes, which in turn depends on the price elasticities of supply and demand for electric cars. If demand is very inelastic—consumers' preferences for electric cars are relatively insensitive to price—consumers will gain from the subsidy but the number of electric cars we see on the road won't change much, nor will CO_2 emissions. And if supply is very inelastic (because production capacity is constrained), companies will gain from the subsidy, but again, the number of electric cars and CO_2 emissions won't change much. Put differently, if demand or supply is very inelastic, it will take a large subsidy to have much of an impact on CO_2 emissions.

A carbon tax also might not do much to change the number of electric cars on the road. The tax will raise the price of gasoline, which raises the cost of owning and operating a gas-powered car, but will shift consumers over to electric cars only to the extent that they are price sensitive. But an electric power producer that currently burns coal but is considering switching to wind

[1] In a competitive market, the incidence of a tax or subsidy depends on the price elasticities of demand and supply. If demand is much more (less) elastic than supply, most of the burden of tax or gain from a subsidy will go to consumers (producers). A good microeconomics textbook will explain why. It's hard for me to think of a better one than Pindyck and Rubinfeld (2018). Buy a copy and read Chapter 9.

power will be very sensitive to the relative prices of coal versus wind, and will be likely to switch in response to a subsidy that lowers the cost of wind, or a carbon tax that raises the cost of coal.

You might say that we can still use subsidies to avoid a carbon tax if we simply identify and target those things that are likely to be most responsive to the subsidy (say, wind power instead of electric cars in my example). Unfortunately that's not easy to do. What is likely to happen instead is that the subsidies will disproportionately go to those firms and industries that have political influence. After all, politicians will decide what to subsidize, and for them economic efficiency is not of primary importance. With a carbon tax we don't have to identify and target anything. The tax will increase the cost of burning carbon in whatever form, and thereby reduce the amount of carbon burned, which is all we really care about.

6.1.3 Government Mandates

Another policy option is direct regulation, i.e., government mandates to reduce fossil fuel consumption in specific ways. For example, the government could require all new electric power plants to run on renewable energy (such as wind, solar, and hydro) rather than fossil fuels. Or it could require all new homes and buildings to be heated by electricity rather than natural gas or fuel oil. Or the government could ban the sale of gasoline-powered cars, in order to accelerate the movement to electric vehicles. In fact, a number of countries have already implemented, or plan to implement, bans on the sale of new gasoline-powered cars.[2]

Requiring electric power to be generated by wind or solar, or requiring new homes and buildings to be heated by electricity, or requiring the sale of all new vehicles to be electric, would indeed reduce our use of fossil fuels, and thus CO_2 emissions. Might this approach to emission reductions be preferable to a carbon tax?

The problem is that the cost of complying with a government mandate depends on the specific mandate, and could be very high. Using solar and wind to produce much of our electricity could probably be done at a moderate cost, but requiring *all* electricity to be produced by solar and wind would be very costly, because it would require a huge amount of battery or other storage

---

[2] Norway has the most ambitious plan; it will phase out sales of gasoline-powered cars by 2025. China, Iceland, Ireland, the Netherlands, Sweden, and the U.K. plan to phase out sales of new gasoline-powered cars by 2030, and Canada, France, and Spain plan to do the same by 2040 (and the state of California by 2035). These plans may change, however, depending on the availability and cost of electric cars.

capacity to keep the electricity flowing when there is no sun or wind. Likewise, in cold climates it is much more efficient to heat homes with natural gas rather than electricity.

So we are back to the same problem we have with subsidies. We would have to determine where mandates would be cost-effective, versus where they would impose costs that are disproportionately high relative to the CO_2 emission reductions they would achieve. As with subsidies, that's not easy to do, so that mandates are likely to be used inefficiently. With a carbon tax we don't have to determine the relative cost-effectiveness of alternative mandates. Once again, the tax simply increases the cost of burning carbon, and lets market forces determine how best to reduce the amount of carbon that's burned.

This does not mean that government mandates and subsidies should not be part of the government's tool kit to reduce emissions. At this point it is unrealistic to think that we will adopt a carbon tax sufficiently large to substantially reduce emissions, so any tax will have to be supplemented by mandates and subsidies. But it is important to be aware of the inefficiencies that mandates and subsidies can introduce, and that a carbon tax avoids.[3]

6.1.4 Cap-and-Trade

We can tax a product or activity that imposes and external cost on society, but we can also limit its quantity. That can be done via direct regulations— you are not allowed to throw litter on the road, and firms are not allowed to dump toxic chemicals in a river or stream. More generally, the government can simply specify how much of a pollutant can be emitted, and impose sharp penalties on firms that emit more than what is allowed.

But there is an efficient way to combine prices with quantity limits. A *cap-and-trade* system to reduce CO_2 emissions would use tradeable emission permits. Under this system, firms would be given a fixed number of permits to emit CO_2, where each permit specifies the number of tons of CO_2 that the firm is allowed to emit. Firms would be heavily penalized for emissions that exceed the amounts allowed by the permits. Permits would be allocated among firms, with the total number of permits chosen to achieve the desired maximum level of CO_2 emissions.

---

[3] Holland, Mansur, and Yates (2020) estimate the inefficiency of a mandate to phase out gas-powered vehicles or a subsidy to encourate the purchase of electric vehicles. They find the inefficiency to be "rather modest: less than 5 percent of total external costs." Holland et al. (2015) and Jacobsen et al. (2020) also discuss these inefficiencies and show how they can be estimated.

A key aspect of this system is that the permits are marketable: They can be bought and sold. This feature makes the system very efficient. Those firms least able to reduce CO_2 emissions would buy permits from firms that can more easily reduce emissions. The total amount of CO_2 emissions would be chosen by the government, but because the permits are marketable, the emission reduction will be achieved at minimum cost.

One disadvantage of a cap-and-trade system relative to a carbon tax is that governments often combine multiple policies. Suppose a country has imposed a limit of 2 Gt of annual CO_2 emissions, and has issued tradeable permits to achieve that goal. Now suppose another policy is added, such as a requirement that all electricity be generated by renewables. The problem is that there will be no additional emissions reduction (because the number of permits that have been issued allows for 2 Gt); there will simply be an increase in cost. With a carbon tax, on the other hand, the added policy will further reduce emissions.[4]

A second disadvantage of cap-and-trade is that it is not obvious how the government can distribute the permits in a fair way, and the distribution process can end up being influenced by political pressures. Ideally the government would auction off the permits, as the European Union is now doing, but in some countries (e.g., the U.S.), companies will resist being forced to pay for permits.

Putting those problems aside, cap-and-trade is an efficient way to reduce emissions. Unfortunately cap-and-trade suffers from the same kind of public resistance that plagues a carbon tax. Limiting emissions will, of course, raise costs for firms that emit CO_2, so they are against it. And many people feel that it is somehow immoral to allow firms to pay to pollute. As a result, despite its efficiency, cap-and-trade for CO_2 has not taken off. One exception is the European Union, which uses a cap-and-trade system—the Emissions Trading Scheme (ETS). The European Commission has announced that it plans to expand the ETS system. But it is unclear to what extent cap-and-trade systems will will become more widely used.[5]

6.1.5 How Large a Carbon Tax?

Now let's come back to a carbon tax. In Chapter 3 we encountered the concept of the Social Cost of Carbon (SCC); it is the cost to society of emitting one

---

[4] Metcalf (2019) discusses this problem in more detail.
[5] For further discussions of cap-and-trade systems and other forms of carbon pricing, see Keohane (2009) and Stavins (2019).

additional ton of CO_2, and is the basis for a carbon tax. So to determine the size of the tax, we just need to calculate the Social Cost of Carbon. Sounds good, but having carefully read Chapter 3, your response might be something like the following: "This is very nice, but we don't know the size of the Social Cost of Carbon. It might be only $30 per ton, or as much as $400 per ton—there is just too much uncertainty over the climate system and over potential climate damages to pin down the SCC. That means we don't know how big the tax should be, and therefore we shouldn't impose a tax."

The first part of your response is correct: "We don't know the size of the SCC." But your conclusion, ". . . therefore we shouldn't impose a tax" is not. We often make personal decisions with very limited information, such as whether to have elective surgery, or whether to get married (to a specific person). And public policy is almost always based on uncertainty over the outcome, such as whether to raise or lower interest rates.

But here's the main reason your conclusion ("we shouldn't impose a carbon tax") is incorrect. Suppose you are in a canoe paddling downstream, and you are told there *might be* a waterfall somewhere along the way. You are in a hurry to complete your trip, and maybe there is no waterfall, or maybe there is but it's a good distance away. Should you pull over to the side of the river every few hundred yards to see what's ahead? Yes, that will slow you down, and you may miss dinner. But it's better than going over a waterfall.

This is the insurance value of climate policy, which I discussed at length in Section 4.1.3. I explained that because of the uncertainties over climate change—the very fact that we don't know the SCC—society should be willing to sacrifice a significant amount of GDP to avoid, or at least reduce, the risk of an extremely bad climate outcome, what I called a catastrophic outcome. The risk of a catastrophic outcome—what is sometimes referred to as "tail risk"—could drive us to impose a carbon tax now, rather than waiting to see how bad climate change turns out to be. In effect, by reducing CO_2 emissions now we would be buying insurance, and the value of that insurance could be considerable.

So, now you're convinced that a carbon tax makes sense. (Wishful thinking on my part, right?) But then the question is how large should the tax be? In the U.S., *any tax* would be better than what we have now. Over the past several decades the U.S. government has been *subsidizing* oil and gas production, which over the past decade has provided fossil fuel producers a benefit of $62 billion per year.[6] The U.S. is not alone; other countries, in fact most other

---

[6] This estimate for the U.S. is from Kotchen (2021). He also calculates the total external cost (including climate, health, and transportation) of these subsidies at just under $600 billion per year.

countries, have also been subsidizing oil, natural gas, and coal, most notably China, Russia, and India, but also the European Union.[7] So even if imposing a carbon tax is not in the cards right now, at the very least we should eliminate the subsidies that are encouraging fossil fuel production.

But we shouldn't give up on a carbon tax. Litterman (2013) and Pindyck (2013c) have argued that given the difficulty of reaching a consensus on the SCC, we should simply impose a modest carbon tax, the exact size of which is not very important.[8] This would at least make it clear to politicians and the public that there is indeed a positive external cost of burning carbon that must be added to the private cost. Later the tax could be adjusted as our understanding of the SCC improves.

I have explained how a carbon tax is a more efficient than government subsidies or mandates as a way to reduce CO_2 emissions. But a carbon tax has another advantage, as explained below.

6.1.6 An International Agreement

The other argument for relying on a carbon tax to reduce emissions is that it makes an international agreement easier to achieve, verify, and enforce. Why? Remember that the Paris Climate Agreement, and for that matter all major international climate agreements, have been based on pledges to reduce emissions by some amount. Much of the negotiation is over how much each country should reduce emissions, and how that amount should compare to the reductions to be made by other countries. India, for example, would (and did) argue that its percentage reduction should be much smaller than that of the U.S. or Europe because it is less wealthy, making its cost of emissions reductions more of a burden.

But there are problems with this approach, the first of which is determining the size of each country's percentage reduction in emissions. As a matter of negotiating tactics, less wealthy countries will argue that their percentage reductions should be smaller than those of wealthy countries. But apart from tactics, how can we actually decide what reductions are indeed fair? How much of a break should poorer countries get relative to their wealthier

---

[7] Coady et al. (2019) have estimated the fossil fuel subsidies for 191 countries, and claim that globally, subsidies in 2015 were on the order of $4.7 trillion (6.3 percent of global GDP). They estimate that without these subsidies, "global CO_2 emissions would have been 28 percent lower."

[8] Rafaty, Dolphin, and Pretis (2020) have shown that simply introducing a carbon tax—of any size— reduces CO_2 emissions, perhaps by making people more aware of the damage those emissions cause. On the other hand, they find that the elasticity is disappointingly small; a $10 per ton tax would reduce emissions by only 0.1 percent.

counterparts? And to what extent should required reductions depend on the starting level of emissions, or on the previous growth rate of emissions? Should we try to equalize emissions per capita? China has about double the CO_2 emissions of the U.S., but about half the CO_2 emissions per capita, so, even putting aside differences in wealth and incomes, should the percentage reduction for the U.S. be more than for China? There are no easy answers to these questions, which greatly complicates the problem of coming to an agreement on country-by-country emission reductions.

An agreement could simply specify *targets* for country-by-country emission reductions as opposed to *required* reductions. (This was the case with the Paris Agreement.) But such an agreement can't be counted on for much because targets need not be met. A stronger agreement, more likely to result in planned aggregate emission reductions, would impose required reductions for each country. That raises the second problem, which is verification. Suppose an agreement is somehow reached that specifies emission reductions for each country. How would we know whether countries are adhering to the agreement? Our data on country-by-country CO_2 emissions comes largely from statistics assembled by each country's government, and governments would have an incentive to overstate their emission reductions.

The third problem is enforcement; what would be done if a country doesn't meet its emission reduction commitment? Without some kind of enforcement mechanism the free-rider problem would kick in: Countries would have an incentive to reduce emissions by less than they pledged.

A tax-based agreement can help get around these problems. Suppose we could come up with a rough consensus estimate of the Social Cost of Carbon on a worldwide basis (i.e., based on climate damages for the entire world, as opposed to the U.S. or any other single country). Becauses it is global in nature, that number would let us determine the carbon tax that should be applied to *all* countries, what we would call a "harmonized" carbon tax. A harmonized carbon tax of this sort could be a superior policy instrument, because it can better facilitate an international climate agreement.[9]

Why would a harmonized carbon tax be preferable to the country-by-country emission reductions that have been the foundation of ongoing climate negotiations? First and foremost, the negotiations would be over a single number—the size of the tax—as opposed to the much more complex problem of negotiating emission reductions for each and every country. It should be much easier for countries with different interests, and different per capita

---

[9] Here, I briefly summarize the argument for a harmonized carbon tax. For more detail, see Weitzman (2014*a*, 2015, 2017) and Pindyck (2017*a*).

incomes and emission levels, to agree to a single number as opposed to a large set of numbers. With country-by-country emission reductions, each country has the free-rider incentive to minimize its own reductions and maximize the reductions of other countries. Of course small countries would still have a free-rider incentive to refuse to take part in a carbon tax regime (as Chen and Zeckhauser (2018) emphasize), but as long as most of the larger GHG emitters do take part, the overall objective of the agreement can still be achieved.

Second, it is difficult to monitor each country's compliance with its agreed-upon emission reductions, and even more difficult to penalize a country that does not comply. A harmonized carbon tax goes a long way towards solving the monitoring problem; compared to emission levels, it is much easier to observe whether countries are indeed imposing the tax to which they agreed. And how can we penalize countries that do not comply? In his paper on "Climate Clubs," Nordhaus (2015) has suggested the imposition of trade sanctions against non-participating or non-complying countries as a way of countering the free-rider problem. While this might indeed increase compliance, it would also risk escalation into a trade war (and involve major modifications to established trade agreements). But once again, as long as the larger GHG emitters join and comply with the tax agreement, the objectives will be largely achieved.

Third, a tax arising out of an international agreement can be politically attractive, making both agreement and compliance more likely. The tax would be collected by the government of each country, and could be spent in whatever way that government wants. Thus it enables a government to raise revenue at a lower political cost. Taxes of any kind are unpopular in much of the world, but in this case politicians can justify the tax burden by saying "the devil made me do it." Finally, an agreement over a harmonized carbon tax can be quite flexible; for example, it need not prevent monetary transfers from rich countries to poor ones, or other forms of side payments.[10]

Targets Versus an SCC-Based Tax

Whether the focus of climate negotiations shifts to a carbon tax, or remains anchored to an agreement over an *equivalent* reduction in total worldwide emissions (which then requires the more difficult agreement over allocating that total reduction across countries), we need a consensus estimate of the SCC in order to come up with the correct tax or emission reduction. As I

---

[10] See Weitzman (2014a) for a detailed discussion of these and other aspects of a harmonized carbon tax. Also, see Kotchen (2018) for a discussion of the use of a worldwide SCC versus a domestic (national) SCC.

have stressed, despite all the research on climate change, we don't have a consensus estimate of the SCC because of all the uncertainties over the climate system and over potential damages from climate change itself. As a results, over the past decade or two, international climate negotiations have focused on *intermediate targets*.

As opposed to "final" targets for emission reductions, these intermediate targets put a limit on the end-of-century temperature increase, which is then translated into limits on the mid- and end-of-century atmospheric CO_2 concentrations, which in turn are translated into required aggregate emission reductions now and in the coming decades. The targeted temperature increase has been generally specified to be 2°C, on the grounds that warming beyond 2°C would take us outside the realm of temperatures ever observed on the planet, and thus could be catastrophic. Some have argued that the correct target should be lower, no more than 1.5°C, although many analyses indicate that even the 2°C limit is probably infeasible given the current atmospheric CO_2 concentration, current emission levels, and plausible assumptions about possible reductions in emissions during the next two decades.

A limit on the end-of-century temperature increase would seem to obviate the need for agreement over the SCC, but in fact it simply replaces the SCC with an arbitrary target that need not have much in the way of an economic justification. Although a temperature increase above 2°C may indeed go beyond anything we have observed, we know very little about its potential impact. Because warming would occur slowly, allowing time for adaptation, there is little reason to conclude that the impact would be catastrophic.

A Temperature Target

Might a temperature target of 2°C make sense? The problem is that without a good estimate of the "damage function," i.e., the loss of GDP that would result from different amounts of warming, there is no reason to think that 2°C is more justified than some other number. Of course if one believed that the true damage function is essentially flat up to 2°C and then jumps dramatically to a level we would consider catastrophic, the 2°C target might indeed make sense.[11] But there is no good reason to believe that there is such a tipping point, or if there is, it would occur at 2°C. (In fact, damage function calibrations in the more widely used integrated assessment models take the GDP loss from a 2°C temperature increase to be less than 3 percent, which we could hardly call catastrophic.)

---

[11] But then what happens after 2100? A global mean temperature that has risen to 2°C by 2100 might be expected to keep rising beyond 2100, so that the 2°C limit for 2100 would be too high.

So why is an essentially arbitrary temperature target the focus of policy? Because it is something that people can agree on, without having to debate the nature of damages (and the extent of adaptation that would likely limit those damages), never mind the discount rate that should be applied to benefits and costs over horizons of 50 to 100 years. Whether or not an agreed-upon end-of-century temperature target can be justified on economic or climate science grounds, it provides a basis for agreement on atmospheric CO_2 concentration targets and thus targets for overall emission reductions.

Is a temperature target of this sort the best we can do? Given the difficulty of estimating the SCC, should the SCC be abandoned as the foundation for climate policy design? If the objective is to *do something* about climate change, then a temperature target might make sense. In fact, we may have reached a point where simply doing something is not entirely unreasonable, even if it is not very satisfying for an economist.

6.1.7 Research & Development

I have argued above that government subsidies for electric vehicles, "green energy," etc., can help reduce CO_2 emissions, but are inefficient relative to the use of a carbon tax. There is one area, however, where subsidies make considerable sense, and that is for research and development (R&D). Why subsidize R&D but not solar panels?

Burning a ton of carbon results in an external cost—a negative externality—which is the basis for a carbon tax. With a carbon tax, you have to pay for the damage you cause society by burning that ton (or part of a ton) of carbon. But if a firm spends money on R&D, it results in a *benefit* for society—a *positive* externality. The reason is that R&D results in new ideas and new knowledge that didn't exist before. Those ideas and knowledge tend to spread out through the economy and benefit other firms that can use them to create new products or reduce the cost of producing its existing products. The firm that originally did the R&D and thereby generated the new ideas and knowledge benefits (the firm might create its own new products), but it can't keep the ideas and knowledge completely to itself. Even if the firm patents some or all of its discoveries, the ideas and knowledge it created will to some extent diffuse throughout the economy and help other firms create products and/or reduce costs.

Because a firm that does R&D cannot completely capture the benefits when the R&D is successful, it will tend to under-invest in R&D, i.e., it will do less R&D than what is socially optimal. That's the basis for the subsidy. With the subsidy, the firm will do more R&D, from which the firm will benefit, but so will society.

And then there is "basic research," that can develop fundamental new ideas that might not directly lead to new products, but can improve our understanding of the science that underlies most R&D. Very little of what comes out of basic research can be captured by a firm that does it, so even with subsidies not enough research will be done. Here the solution is direct funding from the government for work undertaken at research universities, national laboratories, and other research centers.

R&D is especially important when it comes to climate change. We need to find ways to reduce the cost of producing energy without fossil fuels. R&D has already resulted in substantial reductions in the cost of solar and wind power. But solar and wind only produce electricity when the sun is out or it's windy. That means we need to develop better technologies for storing energy, i.e., better battery technologies. The lithium-ion batteries that are so widely used today were first introduced during the 1980s. While research has resulted in extended lifetimes, greater energy density, improved safety, increased charging speed, and lower manufacturing costs, lithium-ion batteries remain an expensive way to store energy. Developing new and better ways to store energy is an example of R&D that today benefits from subsidies to private firms, but also direct funding to universities and other research centers.

6.2 Nuclear Power

A large part of the world's CO_2 emissions (about a third in 2020) comes from the generation of electricity. Furthermore, most scenarios for reducing economy-wide CO_2 emissions involve a widespread substitution of electricity for fossil fuels. So given its importance, how can we "decarbonize" electricity generation? Moving from coal to natural gas will help, and moving to renewables such as wind, solar, and hydropower will help even more. But it is very unlikely that we will reach a point where *all* electricity is generated by renewables—not in the U.S. and Europe, and certainly not in China, India, Russia, and a range of other countries. In that case, what alternatives can we turn to? One alternative is nuclear power.

So far I have said almost nothing about nuclear power, but it may be critical to the decarbonization of electricity production. About 10 percent of the world's electricity is currently generated by nuclear power, with about 440 reactors operating, another 55 under construction, and 109 reactors in the planning stage. However, some 30 to 50 reactors are likely to be decommissioned (shut down) in the coming several years.[12] The use of nuclear power varies widely across countries; it accounts for about 20 percent of electricity generation in the U.S., 70 percent in France, but only about 3 percent in India. (In terms of the total quantity of electricity generated by nuclear, the U.S. is by far the leader with over 800 billion kWh in 2019, compared to 382 billion kWh in France and 330 billion kWh in China.) As Figure 6.1 shows, there has been almost no growth over the past 20 years in nuclear power generating capacity, even though total worldwide electricity production increased by some 75 percent. And projections for the coming decade also show little or no growth in nuclear power.

There are strong objections to the construction and operation of nuclear power plants on the part of the public, and among some environmentalists. Sadly, this aversion to nuclear power has led to an increase in the consumption of coal. An example of this is Germany, which responded to the Fukushima disaster by deciding to phase out all of its nuclear power plants. In 2010, nuclear power accounted for over 22 percent of Germany's electric power generation; that percentage is now below 10 percent and falling. What replaced nuclear power in Germany? Largely coal.[13]

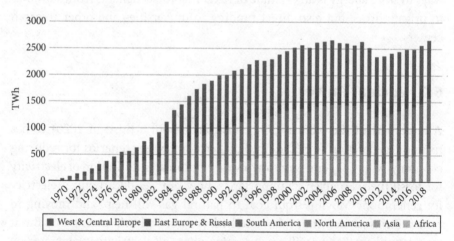

Fig. 6.1 Nuclear Power Generating Capacity by Region. The use of nuclear power peaked by around 2002, and since then has stagnated, even though electricity generation kept increasing. The drop in 2011–2012 is a result of the 2011 Fukushima disaster.

[12] For an overview of the use of nuclear power worldwide, see https://www.world-nuclear.org/information-library/current-and-future-generation/nuclear-power-in-the-world-today.aspx.

[13] The cost of Germany's decision to phase out nuclear power, and replace it partly with coal, is estimated in Jarvis, Deschenes, and Jha (2019).

Nuclear power produces no CO_2 or other GHGs, so it seems like a natural way to "decarbonize" electricity generation and improve overall energy efficiency, i.e., to reduce the amount of CO_2 emitted per quad of energy consumed. Given the importance of reducing CO_2 emissions, why haven't we seen growth in the use of nuclear power? Because while it doesn't produce any CO_2, for many people it produces something else that is problematic: fear. Many believe that nuclear power is inherently dangerous, and fear the construction and operation of nuclear power plants—especially if the plant is built within 10 or 20 miles of where they live. On what is this fear based? First, the very word "nuclear" unsettles people, who associate the word with harmful radiation, or even nuclear weapons.[14] Many people simply don't believe it when they are told that no radiation is emitted by a nuclear power plant.

Even if people came to understand that nuclear power plants don't emit harmful radiation, many would still object to their construction and operation. Why? There are several factors that are at play and have impeded the growth of nuclear power. The most important are the following:

- **A Major Accident.** Much of the public is afraid of a major accident, perhaps a meltdown or explosion that would spew radioactive material over a large area. Surely you, like most people, know about the accidents that occurred at Three Mile Island in the U.S. in 1979 and Chernobyl in 1986, along with the damage to the Fukushima reactors from the 2011 tsunami. Those disasters loom large in people's minds, and have shaken public confidence in nuclear power. People worry that the enriched uranium in a power plant could explode just like an atomic bomb, but they ignore the fact that the uranium has been enriched to only 3 percent to 5 percent U_{235}, not the 90 percent U_{235} needed for a weapon.[15] As for Chernobyl, it happened a long time ago and used an antiquated design. By now, nuclear power technology has advanced to the point where the chance of such a disaster happening again is exceedingly low.

---

[14] Have you ever had an MRI done of some part of your body? MRI stands for *magnetic resonance imaging*, and works by applying a strong magnetic field to the part of the body, which brings the nuclei of the atoms in the tissue under study into alignment. Then strong radio pulses are applied, which force the nuclei out of alignment. As they return into alignment, the nuclei emit electromagnetic signals (i.e. resonate) which can be imaged. When the technology was developed, it was (appropriately) called *nuclear magnetic resonance imaging*, or NMRI for short. But it turned out that the word "nuclear" frightened people, so the technology was renamed, and NMRI became MRI.

[15] U_{235} is the radioactive isotope of uranium. Naturally occurring uranium is only 0.7 percent U_{235} (the rest is the non-radioactive isotope U_{238}). Enrichment works by converting the uranium oxide ("yellowcake") produced from the uranium ore that is mined into uranium fluoride, a gas, which is fed into high-speed centrifuges. The rapid spinning causes the heavier U_{238} to separate from the U_{235}.

Nonetheless, the images of those disasters have generated a perception of great risk—a perception that while unrealistic, has created considerable public opposition to nuclear power. As I'll explain below, the risk of death from nuclear power is far, far lower than from fossil fuels.

- **Nuclear Waste Disposal.** Nuclear fuel rods contain uranium that has been enriched to about 3 percent to 5 percent U_{235}, but eventually the U_{235} gets depleted, falling to a concentration below 1 percent. However, the "spent" fuel rods that are removed from the nuclear reactor contain a variety of other radioactive materials, most notably an isotope of plutonium (Pu_{239}), which can remain radioactive for thousands of years. What to do with this nuclear waste? The spent fuel rods can be temporarily stored, but they must eventually be permanently disposed of. Fortunately, that can be done and done safely. The procedure is to store the spent fuel for up to 40 or 50 years, so that the level of radioactivity has decayed to relatively low levels. Next, the material is sealed inside corrosion-resistant containers (typically stainless steel), which are then buried deep underground in stable geologic rock formations. But there is one problem: Who will pay to develop the needed stable rock formations, and where will they be? (How about in your backyard?) This is particularly a problem in the United States, where tens of thousands of tons of spent fuel are piling up, because there is no political consensus regarding the cost and locations of disposal sites. But like much of climate change policy, this is a political problem, and one that can be readily solved. (How can it be solved? Read the Report to the Secretary of Energy by Blue Ribbon Commission on America's Nuclear Future (2012).)

- **Nuclear Proliferation.** Another concern with a growing use of nuclear power is the possibility that it could lead to nuclear proliferation, which in turn could increase the chances of nuclear terrorism or even nuclear war. There are two perceived channels to proliferation. (1) First, once a country (say, Iran) is in the business of enriching uranium, why stop at the 5 percent U_{235} needed for fuel rods? Why not let the centrifuges keep spinning until reaching the concentration of 90 percent U_{235} needed for a weapon? The answer is economic and other sanctions to force countries with an enrichment capability to agree to limit the enrichment concentration and submit to inspections by the International Atomic Energy Agency. (Again, think Iran.) (2) The second channel is by "reprocessing" the material in spent fuel rods to separate out the plutonium, which can be further processed for use in nuclear weapons. But reprocessing is heavily guarded in the countries where it is done, and it is much more difficult to build a weapon with plutonium versus highly enriched uranium. Furthermore, a country that wanted to build a weapon could do so far more easily by simply enriching uranium or building a reactor designed to produce plutonium; there would be no need for nuclear power plants. Nuclear proliferation will remain a serious threat, with or without nuclear power to produce electricity.

• **The Cost of Nuclear Power.** Most of the cost of nuclear power is up front, i.e., the cost of building the plant. Compared to a coal- or gas-fired plant with the same generating capacity, the capital cost of building a nuclear plant is much greater. The operating cost, on the other hand, is much lower. How much lower, or put more broadly, how does the cost of nuclear power compare to the cost of electricity generation from coal, oil, or natural gas? The answer depends in part on the prices of coal, oil, and natural gas, and those prices fluctuate considerably. Whatever the prices of fossil fuels, a carbon tax would make their effective prices higher. Even without a carbon tax, nuclear power could benefit by reducing the regulatory uncertainty that currently exists. In the United States and other countries, the regulatory framework is unsettled, which has the effect of creating uncertainty over future construction and operating costs, which in turn creates a disincentive to invest in plant construction.[16] Reducing regulatory uncertainty and imposing a carbon tax would go a long way to making nuclear power cost-competitive with fossil fuels.

So, what should we conclude? Is nuclear power still too dangerous? Yes, generating electricity from fossil fuels emits CO_2, but at least it can't lead to a Three Mile Island, Chernobyl, or Fukushima disaster, and to the loss of life that disasters like that can cause. Wouldn't it be a lot safer to turn away from nuclear power rather than expanding its use; in fact, do what Germany has done and shut down our nuclear power plants? Won't that save lives in the long run?

No it won't. It will do just the opposite. The nuclear accidents that have occurred have been rare, and pale by comparison to the many, many accidents and considerable damage caused by the use of fossil fuels. When it comes to saving lives, nuclear power is much safer than any of the fossil fuel alternatives. You're not convinced? Then take a look at Figure 6.2, which shows fatality rates for alternative methods of producing electricity, in deaths from air pollution and accidents, per tera-watt hour (TWh) of electricity produced.

As Figure 6.2 shows, the use of fossil fuels is far riskier than nuclear power. In terms of lives lost per TWh of electricity produced, coal is around 400 times

---

[16] For a (somewhat dated) analysis of the implications of regulatory uncertainty, see Pindyck (1993).

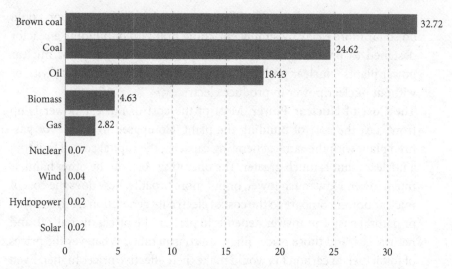

Fig. 6.2 Fatality Risks. Figure compares fatality rates for alternative methods of producing electricity, in deaths from air pollution and accidents, per tera-watt hour (TWh) of electricity produced. The use of fossil fuels is far riskier than nuclear power.
Source: Our World in Data, compiled from Markandya and Wilkinson (2007) and Sovacool et al. (2016).

worse than nuclear, oil is 264 times worse, and natural gas about 40 times worse. Why are fossil fuels so much deadlier than nuclear power? First, the production of fossil fuels, and especially coal, is deadly. Coal miners die from accidents, and (more slowly) from black lung disease, but those deaths don't get the attention that a nuclear accident does. Oil and natural gas production likewise causes accidents. Second, burning fossil fuels generates air pollution; not just CO_2, but *particulates*, tiny particles that can penetrate deep into the lungs, enter the bloodstream, and inflict damage on multiple organs. There is growing evidence that pollution from particulates contributes to infant mortality, and adult mortality from cardiovascular, respiratory, and other types of diseases, and has significantly reduced life expectancy in countries such as China, India, Bangladesh, and Pakistan.[17]

What is the bottom line here? First, using nuclear power to generate electricity is safe, certainly much safer than using fossil fuels. One could argue (as Figure 6.2 shows) that wind, hydro, and solar are even safer, but given the problem of energy storage, it is not realistic to expect *all* electricity will be produced by these renewables. Second, nuclear power is (or could be, with regulatory reform and a carbon tax) cost effective. And third, nuclear power is 100 percent carbon free. The bottom line? We would be making a big mistake

---

[17] For country data, see the report by Health Effects Institute (2020) and references therein.

by dismissing nuclear power as a way of decarbonizing electricity production. There is no question that without nuclear power, reducing CO_2 emissions will be much harder, and for no good reason.

6.3 Removing Carbon

Another approach to deal with the build-up of CO_2 in the atmosphere is to remove some of it (*carbon removal*), and then store it in a way that will prevent its future release into the atmosphere (*carbon sequestration*). Carbon removal and sequestration can help to reduce "net" emissions, or in other words, to "undo" some of the growing CO_2 concentration. Furthermore, in principle it would have no negative environmental impact. Might it provide a realistic solution to the climate change problem? And how, exactly, would it be done?

One obvious way to remove CO_2 is to plant trees, which is indeed seen as a tool for climate policy in some countries. Trees (and other green plants) grow by absorbing CO_2 and combining it with water and the energy from sunlight, releasing oxygen in the process. So, more trees means more absorption of atmospheric CO_2, and lower net emissions. Unfortunately, for the past few decades the world has been cutting down trees—deforestation—at a rapid rate. But suppose deforestation ends and new trees are planted instead. How many trees would have to be planted to make a significant dent in net CO_2 emissions? As I explain below, a very large number.

What about other forms of carbon removal, such as absorbing, sequestering, and storing the CO_2 as it is produced from fossil fuel burning power plants? There have been proposals to do just that, and a number of companies are investing in the development of new technologies. But at this point the technologies are very expensive, currently much too expensive to make economic sense. Might the costs fall in the future, so that some of these technologies will become commercially viable? Could carbon removal and sequestration (CRS) provide a way to significantly reduce net CO_2 emissions? Perhaps, as discussed below.

6.3.1 Trees, Forests, and CO_2

As you learned in high school biology, trees (and other plants) absorb CO_2 and, through photosynthesis, create biomass (wood) and emit oxygen. So if it weren't for the world's trees, the atmospheric concentration of CO_2 would be much higher than it is today.

Sadly, trees are being cut down at a rapid rate. According to the UN's Food and Agriculture Organization (2020), over the five-year period 2015–2020, about 500,000 square kilometers of forest—roughly 1.2 percent of all the world's forests—were cut down. That means fewer trees to absorb CO_2. Even if we completely stopped burning carbon and thereby stopped emitting CO_2, current rates of deforestation would cause the atmospheric CO_2 concentration to increase.

This raises several questions that we need to address. First, how much additional CO_2 is entering the atmosphere as a result of ongoing deforestation? Stated more optimistically, by how much would current net CO_2 emissions fall if deforestation stopped? Second, suppose we plant new trees, either to replace those that were cut down (reforestation), or to create new forested areas (afforestation). By how much would that reduce net CO_2 emissions? How many trees would it take to reduce net CO_2 emissions by 1 Gt per year? And finally, are trees the solution, or at least a significant part of the solution, to our climate problem?

To address these questions, we first need to review a few basic facts and numbers about trees, forests, and their connection to CO_2:

(1) **Land Areas.** It takes land to grow trees, and the common unit of measurement for land areas is the *hectare*. There are 2.47 acres in 1 hectare, and 100 hectares makes 1 square kilometer (1 km^2).

(2) **The World's Forests.** The United Nations' Food and Agriculture Organization (2020) estimates that the Earth's total forest area in 2020 was about 4 billion hectares (or 40 million km^2). The Amazon rainforest accounts for about 530 million hectares, or about 13 percent of the total.

(3) **Number of Trees.** How many trees can be planted in a hectare? As you'd expect, it depends on the type of trees and the climate (temperature and precipitation, and their variability). Crowther et al. (2015) have estimated the global number of trees to be about 3 trillion, which would imply a density of $(3 \times 10^{12} \text{ trees})/(4 \times 10^9 \text{ hectares}) = 750$ trees per hectare. Ter Steege et al. (2013) have estimated a lower number for the Amazon rainforest, namely 565 trees per hectare. However, other estimates put the average density in a range from 1000 to 2500 trees per hectare. (See, e.g., http://nhsforest.org.) In what follows, I will use the lower end of this range, i.e., a density of 1000 trees per hectare.

(4) **Absorption of CO_2.** How much CO_2 is absorbed by a tree each year on average? The answer depends on the type, size and age of the tree, and the density of the forest. (Trees planted closely together absorb

less carbon than those planted in the open where they can grow larger.) According to the European Environmental Agency, a mature hardwood tree absorbs about 22 kg of CO_2 per year.[18] Not all trees are mature, but 20 kg per year is a reasonable average number for annual CO_2 absorption per tree. With 1000 trees per hectare, $1000 \times 20 = 20,000$ kg $= 20$ tons of CO_2 is absorbed per year from each forested hectare.

With these numbers we can address the questions raised above: (1) By how much would net CO_2 emissions fall if current rates of deforestation ceased? (2) If we plant new trees, to replace those cut down (reforestation) or to create new forested areas (afforestation), how many trees would it take to reduce net CO_2 emissions by 1 Gt per year? (3) Could planting trees be a significant part of the solution to our climate problem?

Deforestation

Deforestation has been a problem for a long time, as large tracts of land have been cleared for farming, ranching, and timber production. There has been particular concern about the loss of trees in the Amazon rainforest, and the possibility that a "tipping point" could soon be reached at which the loss becomes permanent. Deforestation has also occurred at a rapid rate in Indonesia and Malaysia, mostly to clear land for palm oil production, and in those countries most of the trees that were cut down were burned, releasing large amounts of CO_2 as well as pollution from smoke and particulates.

Deforestation can cause many problems—for example, changes in local/regional rainfall patterns (e.g., longer drier seasons); soil erosion and the resulting loss of arable land; an increased threat of flooding; the loss of animal and plant habitat, sometimes to the point of driving species to extinction; biodiversity loss; and increased emissions of greenhouse gases. The focus here, however, will be limited to the effects of deforestation on CO_2 emissions and thus climate change.

Rates of deforestation have actually declined over the past decade or two. Figure 6.3 shows annual rates of deforestation in the Brazilian part of the Amazon rainforest, which accounts for about 60 percent of the total rainforest, i.e., about 320 million hectares out of a total of 530 million hectares. As the figure shows, deforestation peaked in 2004 and then fell, but started rising again in 2015. The average rate of deforestation in the Brazilian Amazon over the period 2000–2019 (the horizontal line in the figure) was about 1.2

[18] See https://www.eea.europa.eu/articles/forests-health-and-climate-change/key-facts/.

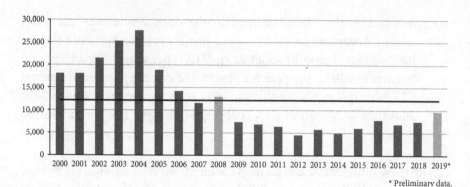

Fig. 6.3 Annual Deforestation Rates in the Brazilian Amazon (km^2 per year, so multiply by 100 to convert to hectares). Brazil accounts for about 60 percent of the total Amazon rainforest, which in turn accounts for about 13 percent of Earth's total forest area.
Source: Amazon Fund (2019).

million hectares (12,000 km^2) per year, or about 0.37 percent of a total forest area of 320 million hectares. But for 2015–2019 deforestation was about 0.8 million hectares per year, i.e., 0.25 percent of the total forest area. This is about the same as the recent global rate of deforestation; the United Nations' Food and Agriculture Organization (2020) estimated global deforestation from 2015–2020 at around 10 million hectares per year, or about 0.25 percent of the total global forest area.

Does 10 million hectares per year of deforestation imply a loss of 10 million hectares worth of trees each year? No, because that loss is partially offset by new trees, some of which result from forest regrowth in abandoned agricultural land, and some that are planted each year for consumption, either as timber or for firewood. These new trees replaced some of those cut down, and also created new forest areas. What matters is the *net loss* of trees. As shown in Figure 6.4, during the period 2015 to 2020, this was about 6 million hectares per year—about 10 million hectares per year of deforestation was partly offset by about 4 million hectares per year of forest gain.

Figure 6.4 also shows how the rate of deforestation and the annual net forest loss has changed over time. Note that net forest loss was greatest—almost 8 million hectares per year—during the 1990s. Annual net forest loss declined substantially through 2015, but increased again during 2015–2020 to about 6 million hectares, largely because of a reduction in forest gain.

What will happen to net forest loss over the coming decades? Recently deforestation has accelerated in Brazil, Indonesia, and Malaysia, but that could change in response to new government policies. d'Annunzio et al. (2015) developed a detailed model of regional forest losses and gains, and

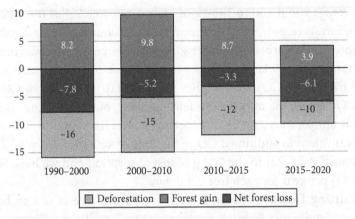

Fig. 6.4 Net Annual Loss of Forests. For each period of time, the figure shows estimates of annual global deforestation, annual global gains in forest areas, and the net loss, all in millions of hectares. During 2015–2020 annual deforestation was about 10 million hectares, but this was partially offset by 3.9 million hectares of forest gain, for a net annual loss of 6.1 million hectares.

Source: United Nations' Food and Agriculture Organization (2020) and https://rainforests.mongabay.com/deforestation/.

projected a deceleration of the rate of net forest loss over the decade 2020–2030. But these (and related) projections are subject to considerable uncertainty. What will happen? We don't know. But we need to consider how forest losses (or gains) will affect CO_2 emissions, and thus affect climate.

Deforestation and CO_2

To understand what deforestation (and reforestation) does to CO_2 emissions, we need to review a few more facts and numbers:

(1) **Cutting Down a Tree.** What happens to net CO_2 emissions when a tree is cut down? On average, a tree absorbs about 20 kg of CO_2 per year (see page 159), so you might say that the loss of the tree would cause net annual CO_2 emissions to increase by 20 kg. But that's not right—net CO_2 emissions would increase by more than 20 kg per year. Why? Because the tree has accumulated a large amount of carbon over its lifetime, which is now in the form of wood. If the tree is burned (which is usually what happens), or simply falls to the ground and decays over time, that accumulated carbon is oxidized and emitted as CO_2. How much CO_2? That depends on the type, size, and age of the tree. For tropical forests like the Amazon, there is on

average about 130 to 145 kg of carbon per tree.[19] But for other types of forests (e.g., tropical or temperate deciduous forests, and coniferous forests) the carbon content is somewhat lower. A conservative average number is 110 kg of carbon per tree. Since one ton of carbon yields 3.67 tons of CO_2, this implies $(3.67) \times (110) \approx 400$ kg of CO_2. That CO_2 enters the atmosphere fairly quickly, but to measure its impact on temperature, we can "amortize" it over 10 years, so it is roughly equivalent to additional CO_2 emissions of $400/10 = 40$ kg of CO_2 per year.[20] Add that to the 20 kg of lost absorption and we have 60 kg of CO_2 per year for each tree cut down.

(2) **Cutting Down a Forest.** Cutting down one tree is not so bad, but cutting down a forest is another matter. Recall from Figure 6.4 that over the five-year period 2015–2020, about 10 million hectares of forest were cut down each year, offset by about 4 million hectares of forest gain, for a net loss of 6 million hectares. With 1,000 trees per hectare, that corresponds to net loss of 6 billion trees per year. With 60 kg of CO_2 per year for each tree cut down, the net annual loss of forest has added roughly 60 kg $\times$ 6 billion = 360 billion kg, or 0.36 Gt, to our annual net emissions of CO_2.

(3) **Cutting Down a Forest Each Year.** If we have a net loss of 6 million hectares of forest in one year *but only one year*, annual CO_2 emissions over the next 10 years would be about 0.36 Gt higher than it would be otherwise. Not good but not terrible. The problem is that we have been losing around 6 million hectares of forest *every year*. That means we are increasing net CO_2 emissions by an additional 0.36 Gt *each year*. If we maintain that rate of deforestation, over 10 years we would have increased net annual CO_2 emissions by around 3.6 Gt, which is about 10 percent of the recent level of 37 Gt of global CO_2 emissions.

(4) **Overall Impact on CO_2 Emissions.** These rough calculations above show that the net loss of 6 million hectares per year of forest has resulted in annual CO_2 emissions some 3.6 Gt higher than it would be otherwise—roughly 10 percent higher. This is a conservative estimate, as other studies have suggested that tropical deforestation *alone* could

---

[19] More accurately, 130 to 145 tons of carbon per hectare, but I am assuming a density of 1000 trees per hectare. See, e.g., Gibbs et al. (2007) and Ramankutty et al. (2007). And remember that one ton of carbon yields 3.67 tons of CO_2.

[20] See Amazon Fund (2010) and Franklin and Pindyck (2018). Remember that an increase in the atmospheric CO_2 concentration affects temperature with a lag of around 20 to 50 years, so injecting a "pulse" of 400 kg of CO_2 into the atmosphere will have roughly the same warming effect as injections of 40 kg per year over 10 years. Also, I am ignoring trees that are farmed to produce lumber that is used in construction; when those trees are cut down the carbon is (at least temporarily) sequestered.

account for some 3.0 to 3.7 Gt of CO_2 emissions.[21] The conclusion: Deforestation has made a significant contribution to global net CO_2 emissions.

Reforestation and Afforestation

A loss of trees adds to net CO_2 emissions, so eliminating that loss and turning it into a gain could reduce net emissions. But by how much? And what would it take to make a significant dent in global net CO_2 emissions?

We have seen that during 2015–2020 the world experienced annual net losses of about 6 billion trees (6 million hectares of forest), and those losses could account for some 3.6 Gt of CO_2 emissions. So if the rate of deforestation could be reduced to a level that is completely offset by forest gains, net emissions would, after a lag, fall by about 3.6 Gt per year. If forest gains during the coming decade would otherwise look the same as during 2015–2020, that 3.6 Gt reduction could be achieved by reducing the rate of deforestation from 10 to 4 billion hectares per year. That's a big drop, but as Figure 6.4 shows, the rate of deforestation has already declined somewhat during the past 20 years, and further declines might well be feasible.

What about planting new trees, as part of either *reforestation* (planting trees in areas that have been partially or totally deforested, i.e., rebuilding what had been existing forests), or *afforestation* (planting trees in areas that were previously not forested, i.e., creating new forests)? Afforestation helps maintain natural forests by providing an alternative source of tree products. Currently it is done mostly for commercial purposes, i.e., because the tree products are in high demand. But it can also be done to absorb CO_2 and reduce net emissions.

The impediment to both reforestation and afforestation is the need for land. Remember that deforestation is mostly done to create areas that can be used for agricultural purposes. This can include grazing for cattle, sheep, and other animals, farming a variety of crops, or palm oil production as in Indonesia and Malaysia. The problem is that reforesting that land would mean giving up its use for agriculture. But some deforestation is done simply to produce timber and wood pulp (in an unfortunately unsustainable way), in which case the land might be easier to reclaim.

Putting cost aside, is there enough land available for afforestation on a scale that could substantially reduce net CO_2 emissions? The answer is unclear.

---

[21] See, e.g., Baccini et al. (2012) and Harris et al. (2012). As mentioned earlier, deforestation can have negative impacts that go beyond CO_2 emissions and climate change; see Watson et al. (2018) for a nice overview.

In one of the more optimistic studies, Bastin et al. (2019) show that close to one billion hectares of land could be turned into forest with about 1000 trees per hectare. If they are right, what would an additional billion hectares of forest do to CO_2 emissions? Quite a bit. If the tree density of the newly forested land is 1000 trees per hectare, and each tree (when mature) can absorb 20 kg of CO_2 per year, the total annual absorption would be (1000 trees) $\times$ (20 kg) $\times$ (10^9 hectares) = 2×10^{13} kg = 20 Gt of CO_2. That's more than half of global CO_2 emissions.

And how many new trees would be needed to reduce annual CO_2 emissions by just 1 Gt? We just saw that 1 billion hectares of new forest would reduce emissions by 20 Gt, so we'd need 50 million hectares of new forest to get a 1 Gt reduction. With 1000 trees per hectare, that comes to 50 billion new trees.

We have ignored the opportunity cost of turning large tracts of land—land that might otherwise be used for a variety of other purposes—into forest, and we have ignored the water that would be needed to support the tree growth. But these calculations show that in principle, afforestation could indeed help to reduce net CO_2 emissions.[22]

Are Trees the Solution?

Yes and no. Yes, because in principle a sharp reduction in the rate of deforestation could reduce net CO_2 emissions by about 10 percent (and have other environmental and ecological benefits as well), and large-scale afforestation (to the tune of a billion hectares) could cut net CO_2 emissions in half. No, first because deforestation has been—and is likely to continue to be—driven by strong economic forces. And second, because turning even a fraction of a billion hectares of land into new forest would be an extremely expensive undertaking, and it is unclear who would pay for it.

So where does that leave us? Trees won't solve our climate problem, but they can help, and should be viewed as a component of climate policy. First, we should reduce the recent rates of deforestation as much as possible. This would reduce net CO_2 emissions somewhat, but also have other important environmental benefits. Second, we should pursue options for afforestation to the extent that those options make economic sense. Even 100 million hectares of new forest (just 10 percent of the billion hectares considered above) would reduce net CO_2 emissions by about 2 Gt, which is more than 5 percent.

Finally, it is important to keep in mind that CO_2 is absorbed not only by trees, but also by other plants, particularly those found in coastal wetlands,

---

[22] Houghton, Byers, and Nassikas (2015) also argue that afforestation could make a signicant contribution to CO_2 removal, but they note that large amounts of water would be need to support the tree growth.

including mangrove forests, tidal marshes, and seagrass meadows. Thus it is critical to preserve existing wetlands, and also to restore and even build up these ecosystems as a way to aid in CO_2 removal.

6.3.2 Carbon Removal and Sequestration

Apart from planting trees, there are other ways to remove carbon from the atmosphere, sequester it, and thereby reduce net CO_2 emissions. At the risk of oversimplification, the two basic approaches to carbon removal and sequestration (CRS) are the use of biomass to generate energy ("bioenergy"), and direct air capture, i.e., pulling CO_2 from the atmosphere or from the exhaust of fossil fuel burning power plants.[23]

For both of these approaches to CRS, a variety of technologies have been or are being developed and tested. Most are currently very expensive, but there is hope that costs will come down once firms move down the learning curve and economies of scale kick in, or as the technologies themselves evolve.

Bioenergy

Trees remove CO_2 from the atmosphere, but so do other plants. So one approach to CRS is to harvest different kinds of plants and algae to produce various sorts of biomass that can be burned. Burning the biomass would of course release CO_2 into the atmosphere, but the *net* emissions would be zero, because the CO_2 was first extracted from the atmosphere by the plants and algae. In this case, burning the biomass just puts back into the atmosphere what the plants took out, with no net change in the CO_2 concentration.

A related approach to CRS is being pursued by the Drax Power Station in Yorkshire, England.[24] The Drax plant was originally the biggest coal-fired power station in Britain, but now it generates electricity with the use of almost no coal. Instead, the plant burns biomass in the form of compressed wood pellets. (In 2019, the biomass burned at Drax accounted for over 10 percent of Britain's renewable energy, about the same fraction as the country's solar panels.)

But isn't there a potential problem with the use of wood pellets as a fuel? The wood comes from trees, so in principle trees would have to be cut down to provide the wood. For now, Drax has gotten around this problem by using

---

[23] For overviews of different approches to CRS, see Smith, Friedmann et al. (2017) and National Research Council (2015), as well as a nice summary in *The Economist*: https://www.economist.com /briefing/2019/12/05/climate-policy-needs-negative-carbon-dioxide-emissions.

[24] See https://www.drax.com/about-us/our-projects/bioenergy-carbon-capture-use-and-storage-beccs/.

sawdust and other refuse from sawmills to make the pellets. However, to generate electricity from wood pellets on a large scale, it is unlikely there would be enough sawmill refuse, which means trees would have to be cut down to produce the wood pellets.

The production of wood pellets (or other forms of wood-based biomass) could still be sustainable if the wood is harvested from a fixed area, in which new trees are continually grown to regenerate the supply of wood. But if wood-based biomass is ever to be used to produce electricity on a large scale, the land required for the planting, harvesting, and regeneration of trees would have to be very large. At this point, it is unclear where the land would come from and how expensive it would be to use it to grow and harvest trees. Alternatively, other naturally regenerating forms of biomass could be used, such as algae or agricultural waste. But again, it is unclear how feasible and costly it would be to harvest such forms of biomass on a large scale.

Air Capture and Exhaust Capture

Another way to reduce net CO_2 emissions is by *air capture*, which means pulling CO_2 directly out of the air ("direct air capture"), or out of the exhaust of coal-burning power plants. Direct air capture is done by using chemical or physical processes to separate CO_2 from the air.[25] Occidental Petroleum has been developing a plant (along with Carbon Engineering, a Canadian firm) in Texas to demonstrate the feasibility of direct air capture. The operation, described in Occidental Petroleum Corporation (2020), would convert the captured CO_2 into oil. Burning that oil would result in zero net emissions (because the CO_2 emitted from burning the oil would be just offset by the CO_2 captured from the air to produce the oil.) But right now this process is very expensive, in part because the concentration of CO_2 in the atmosphere is so low (about 0.04 percent).

Of course the concentration of CO_2 released by a coal-burning power plant is much greater, so it should be easier to capture it from the plant's exhaust, and indeed that was one of the earliest plans for CRS. Here the idea was to use a chemical process to extract the CO_2 from the exhaust, and then pump it deep underground, where it would remain. The Drax Power Station in Yorkshire, England has been developing a version of this technology; the process is described in Figure 6.5, which was reproduced from their website.

While this process looks easy, there are a number of problems that must be solved. First, while it is easier to remove CO_2 from power plant exhaust than it is from the air, at this point it is still very expensive, and only about 90 percent

---

[25] The processes are described in detail in Sanz-Pérez et al. (2016).

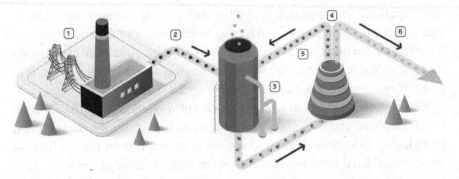

Fig. 6.5 Removing CO_2 from an Emissions Source. The Drax Power Station has been developing a technology for removing CO_2 from the exhaust of a coal-fired power plant. Exhaust containing CO_2 (1) is cooled and treated (2), then fed to an absorption tower (3) which absorbs the CO_2 in a solvent. The solvent is heated in a boiler (4) which separates the CO_2, and the solvent is re-used (5). The CO_2 is then transported via pipeline (6) for permanent storage under the North Sea.
Source: Drax Power Station, https://www.drax.com/about-us/our-projects/bioenergy-carbon-capture
-use-and-storage-beccs/.

efficient, so some CO_2 will still escapte into the atmosphere. Second, pumping the CO_2 into an underground formation where it will remain permanently is also expensive, and will require a large amount of underground capacity if the use of this technology is ever to be widespread. And finally, extracting the CO_2 and pumping it underground requires energy, which would have to come from a zero-carbon source such as wind, solar, or nuclear. As a result, CRS, whether via air capture of power plant exhaust capture, is still far from being economical.

The good news is that a great deal of research is underway on new approaches to air capture, whether directly from the atmosphere or from power plants' exhaust. (For recent analyses of the costs and feasibility of evolving technologies to remove and sequester carbon, see Li et al. (2015), Keith et al. (2018), Krekel et al. (2018), and Ranjan and Herzog (2011).) Some of this research may lead to more cost-efficient approaches to CRS. And even for some of the existing technologies, costs are likely to fall as a result of scale economies.

6.3.3 The Bottom Line

To what extent can carbon removal and sequestration provide a way to reduce net CO_2 emissions and thereby "undo" some of the growing CO_2

concentration? This approach to dealing with climate change is attractive in part because it would let us avoid having to eliminate all CO_2 emissions, and it would have no negative environmental impact itself. And some have argued that it is probably the only way we can come close to limiting the increase in global mean temperature to a 1.5°C or even a 2°C target.[26]

We examined how forests and other forms of vegetation can pull CO_2 from the atmosphere, and how deforestation can add CO_2 to the atmosphere, first by reducing CO_2 absorption, and second as stored carbon (in the form of wood) is oxidized from the burning or the natural decay of trees that have been cut down. We have seen that reducing the recent rates of deforestation can make a significant difference in net emissions, but deforestation is driven by strong economic forces. Reforestation and afforestation can also help, but large numbers of trees and large tracts of land would be needed (along with substantial water resources). Once again, economic forces would have to be overcome; the opportunity cost of using land for trees could be substantial, and it is unclear as to who would pay for large-scale forestation efforts.

Despite these obstacles, trees can help reduce net CO_2 emissions, and should be a component of climate policy. This means reducing the rate of deforestation as much as possible, which would not only reduce net CO_2 emissions somewhat, but would also have other important environmental and ecological benefits. And we need to pursue options for reforestation and afforestation to the extent that those options make economic sense. Finally, CO_2 is also absorbed by other plants, particularly those found in coastal wetlands, so it is important to preserve, restore, and where possible build up these wetlands.

And then there are other forms of carbon removal and sequestration (CRS), such as absorbing and storing the CO_2 produced from fossil fuel burning power plants. A good deal of R&D has been directed at CRS, and a number of companies are investing in the development of new approaches. But at this point the technologies are much too expensive to make economic sense. The costs may indeed fall over time, so that some of these technologies will become commercially viable, and might begin to be used on a large scale in the U.S. and Europe. But even if costs fall, it is unlikely that large-scale adoption of these technologies will occur in China, India, and other countries that are currently large emitters of CO_2. CRS will help, but it isn't something we can count on to get even close to net zero emissions on a global scale.

---

[26] See, e.g., Fuss et al. (2016) and Fuss (2017).

6.4 Further Readings

This chapter focused on the importance of reducing global CO_2 emissions, and why it must be part of an international agreement that can be monitored and enforced. I argued that the most efficient (i.e., least costly) way to achieve this goal is through the use of a carbon tax. But to the extent possible, we should also try to remove CO_2 from the atmosphere, thereby reducing net emissions. The discussion here has been brief, and there is more to read.

- I have argued, as would most economists, that the most efficient way to reduce CO_2 emissions is through the use of a carbon tax. *Paying for Pollution: Why a Carbon Tax is Good for America* by Metcalf (2019) makes the case for a carbon tax clearly and convincingly, and explains how such a tax would be designed and applied. The book also provides a very readable introduction to the economics of climate change.
- Aldy et al. (2010) and Metcalf (2009) provide nice overviews of how CO_2 abatement policies can be designed and implemented, and some of the difficulties involved in reaching an international climate agreement.
- Innovation will be critical to dealing with climate change over the coming decades. We need to find new ways to store energy, to remove carbon from the atmosphere and sequester it, to produce concrete, steel, aluminum, and other materials in far less carbon-intensive ways, . . . the list goes on. For a speculative but inspiring overview of ways that R&D and innovation can help, see Gates (2021), the recent book by Microsoft founder Bill Gates.
- Are you still wary of nuclear power, and even after reading Section 6.2 concerned that it might be a risky way to reduce CO_2 emissions? Then read the report by the International Energy Agency (2019) for a more detailed treatment of the safety and policy issues connected with nuclear power.
- There is a large and growing literature about trees and their connection to CO_2 emissions, as well as the effects of deforestation. See the United Nations' Food and Agriculture Organization (2020) and United Nations Environment Programme (2020) for discussions of the causes, extent, and impacts of deforestation and net forest loss. As mentioned, deforestation can have negative environmental and ecological impacts that go beyond CO_2 emissions and climate change; Watson et al. (2018) provide

a nice overview. For more detail on the Amazon rainforest, see Amazon Fund (2019).

- Our discussion of technologies for carbon removal and sequestration provided only a brief introduction to the topic. For more detailed overviews, see Hepburn et al. (2019), Smith, Friedmann et al. (2017), and National Research Council (2015).

7
What to Do: Adaptation

Reducing net CO_2 emissions—via a carbon tax, subsidies for "green" energy technologies, government mandates to reduce fossil fuel consumption, developing and implementing alternative approaches to carbon removal and sequestration, and pursuing research on energy storage and other technologies to help us shift to renewables—is critical, and should continue to be a fundamental part of climate policy. But at the same time we have to recognize that despite our best efforts, it is unlikely that global CO_2 emissions will be reduced enough to prevent a temperature increase of 1.5 or 2.0°C by the end of the century. That brings us to the next critical component of energy policy—investing in adaptation.

Suppose the weather is sunny and warm, so you plan a relaxed day at the beach. You arrive when it's low tide, so you set a chair down in the sand, and start reading that book you just bought. But now the tide starts to come in, so what will you do? If you just sit and keep reading your book, you'll eventually find yourself under water. So you get up and move your chair farther back on the beach. That's adaptation. You can anticipate that the tide will rise, so you adapt by moving your chair.

We can also anticipate climate change, even though we can't predict what it will entail nearly as accurately as we can predict the tides. In this case, adaptation means taking steps to counter the warming effects of a high and rising CO_2 concentration, or any of the other aspects of climate change that warming can bring about. Adaptation can include developing new crops that can resist extreme temperatures, adopting policies to discourage building in flood-prone or wildfire-prone areas, building sea walls to prevent flooding, and the use of geoengineering to reduce the greenhouse effect of a rising CO_2 concentration.

And adaptation can be private, meaning actions taken by households or firms; public, meaning actions take by local, state, and federal governments; or a mix of the two. Examples of private adaptation include decisions by real estate developers to avoid new construction in coastal areas vulnerable to hurricanes, investments by firms like Carrier Corporation in the development of cheaper and more efficient air conditioners, and decisions by households to

Climate Future: Averting and Adapting to Climate Change. Robert S. Pindyck, Oxford University Press.
© Oxford University Press 2022. DOI: 10.1093/oso/9780197647349.003.0007

install air conditioners. Examples of public adaptation include the construction of flood walls, dikes, levees, and other barriers to prevent flooding from rising sea levels, and various forms of geoengineering. Adaptation can also involve a mix of private and public actions; an example is the development of heat-tolerant strains of wheat, corn, and other grains by private agricultural firms, with the support of research performed and/or funded by the government (the Department of Agriculture in the U.S.). Migration, with or without government involvement, is another example of adaptation; we have seen that people respond to long-term changes in temperature by moving to regions that are cooler on average.[1]

In what follows, I will discuss in detail three areas of adaptation: agriculture (e.g., the development and adoption of alternative crops, new methods of irrigation, and planting in different locations); ways of reducing damage from hurricanes and rising sea levels; and approaches to geoengineering. There are many other forms of adaptation, but this should at least give you the flavor of what is possible—and perhaps necessary.

7.1 Adaptation in Agriculture

As explained earlier in this book, we know very little about how climate change will affect the economy over the long run. If the global mean temperature rises by 3°C over the next 50 years, will GDP be 5 percent lower than it would be otherwise? 10 percent lower? We just don't know, in large part because the warming will occur slowly, allowing for the possibility of natural adaptation. And whatever overall impact climate change will have, it is likely to differ considerably across regions of the country and across sectors of the economy. For example, it is likely to have a minimal impact on the production of consumer electronics, computer software, and pharmaceuticals, but may have a much larger impact on agricultural output, which can be very sensitive to the weather. Temperature extremes (whether hot or cold) and rainfall extremes (too much or too little rain) can sharply reduce crop yields. So it is of no surprise that agriculture is the sector that has been most intensively studied in terms of climate change impacts.

But even in the case of agriculture it is difficult to determine the likely impact of future climate change. The problem is that we can examine how fluctuations in the *weather* have affected crop yields, but that doesn't tell us

---

[1] For example, Mullins and Bharadwaj (2021) have shown that migration across counties in the U.S. responds to long-term variation in temperatures, but not to the short-term variations.

very much about the effects of changes in *climate*. *Climate* describes the kind of weather (and the extent of variations in the weather) that we can expect *on average*, year after year. Miami, for example, has a much warmer and more humid climate than, say, Minneapolis. But today's weather in Minneapolis might be much warmer than usual for this time of year, and might be very different from the weather a few weeks ago.

The weather in any location changes frequently.[2] The climate in any location, if it changes at all, does so very slowly. The implication for agriculture: A change in the weather—say, an unusually hot or cold summer—can affect crops but might have nothing at all to do with climate change.

7.1.1 What Can the Data Tell Us?

So we have a problem. We want to know how climate change will affect agricultural output, but taking into account the potential for adaptation. We have plenty of data on crop yields and on weather (temperature and rainfall) in different locations. So we can look at how crop yields have changed in response to unusually warm or cold weather, but that won't tell us much about how crop yields will change in response to gradual changes in climate.

Economists have tried to get around this problem in essentially two different ways. First, they have tried to compare crop yields in different areas that have different climates. For example, we might compare crop yields in an area of the U.S. that is warmer on average (say, Louisiana) with an area that is cooler on average (say, North or South Dakota). If yields are higher in the cooler area, then perhaps we could conclude that as the cooler area warms up (e.g., the climate in the Dakotas becomes more like the climate in Louisiana), its crop yields will fall. Of course there are many other factors that can affect crop yields in different areas, such as humidity, soil characteristics, etc., but we could try to take those differences into account, and thereby do an "all other things equal" comparison.[3]

A second approach to the problem is to look at the relationship between crop yields and weather over relatively long periods of time (50 years or even more), as opposed to year-to-year changes. Suppose, for example, that temperatures fluctuate from year to year, but over 50 years, the average temperature rose by 1°C, which might resemble something closer to climate

---

[2] Especially here in Boston. Mark Twain is supposed to have said "If you don't like the weather in New England now, just wait a few minutes."

[3] Mendelsohn, Nordhaus, and Shaw (1994) was one of the earliest studies that compared crop yields in different areas with different climates, and tried to account for other factors that could affect yields.

change. If over those 50 years average crop yields fell by some amount, we might conclude that climate change was the cause, and that future increases in the average temperature will cause further declines in crop yields.[4]

Both of these approaches are informative, but still suffer from a fundamental problem: They don't fully account for adaptation. To understand the problem, suppose that Location A has a warmer climate than Location B and lower crop yields on average. Does this mean that if climate change causes Location B to become as warm as Location A, the crop yields in B will drop to match A's current yields? Not necessarily, because any climate change will occur slowly, and thereby give farmers in Location B time to adapt. How might they adapt? Perhaps by changing the types of crops they plant (moving towards crops that are less sensitive to warm weather), or even developing and/or planting new hybrid crops that are robust and less sensitive to temperature.[5]

If adaptation to rising temperatures didn't occur in the past, why should we think it will occur in the future? Because in the past, when climate change was not widely understood to be a real and serious threat, farmers may have seen little reason to spend the money (and other resources) needed to adapt. The key is that adaptation to climate change is more likely to occur when the threat is perceived to be real and imminent. And today, much more than in the past, the threat of climate change is indeed perceived to be real and imminent.

7.1.2 An Historical Experiment

So what can we do to determine how much climate change is likely to affect agriculture, and the extent to which adaptation will limit its impact? Ideally we'd like to run an experiment in which we change the climate and then see what happens. (And then if we don't like what happens, restore the climate to the way it was before.) We can't run that experiment, but perhaps we don't need to: To some extent the experiment has already been run for us.

---

[4] Studies based on time variation over 50 years or more include Deschênes and Greenstone (2007) and Schlenker and Roberts (2009). For an overview of the use of weather data to infer potential climate change impacts on agriculture, see Auffhammer et al. (2013).

[5] In a more recent and innovative paper, Burke and Emerick (2016) exploit the fact that "changes in climate have been large and vary substantially [across counties in the U.S.]; temperatures in some counties fell by 0.5°C between 1980–2000 while rising 1.5°C in other counties, and precipitation across counties has fallen or risen by as much as 40 percent over the same period." Using 20-year "long differences," they estimate how county-level agricultural outcomes responds to these changes in temperature and precipitation. They find that the long-run response is not very different from the short-run response, suggesting that adaptation is limited. But as they acknowledge, "If farmers failed to adapt in the past because they did not recognize the climate was changing, but in the future they become aware of these changes and quickly adapt, then our findings would be a poor guide to future impacts of warming."

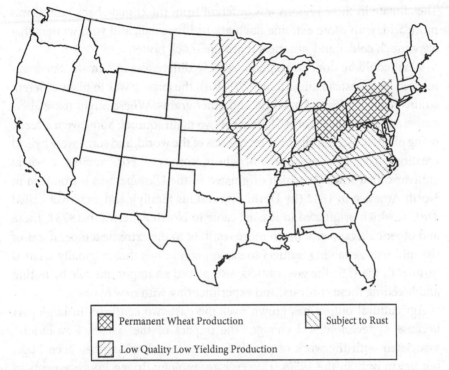

Fig. 7.1 Wheat Production in 1858. During the 1850s, most production took place in the eastern part of the U.S. Later people moved to the mid-west, but wheat yields were low because of the more extreme temperatures. Adaptation took the form of planting new cultivars and developing hybrid grains.
Source: Olmstead and Rhode (2011*b*).

It turns out that history has already provided an interesting and informative experiment on agricultural adaptation to climate change, an experiment described and analyzed by Olmstead and Rhode (2011*b*,*a*). The climate "change" in question did not occur over time, but rather across space, that is, across regions of the United States. Remember that people first settled in the eastern part of the U.S., and then later started moving west. As they moved and settled in what are now mid-western states such as Iowa, Illinois, Missouri, Wisconsin and Minnesota, what did they find? Soil that was much easier to till than the rocky soil in the east, and seemed ideal for planting wheat and corn. So they planted, but the resulting crop yields were poor. Why? Because compared to back east, the climate was harsh.

Figure 7.1, drawn from Olmstead and Rhode (2011*b*), illustrates how during the 1850s, most wheat production in the United States took place in New York, Pennsylvania, and Ohio. People moved west, settled, and tried to farm the much richer soil that they had found—but with very limited success.

The climate in these regions was different from the climate back east. It was more arid, with more extreme fluctuations in rainfall, and with winters that were much colder and summers that were much hotter.

What could be done to respond to these differences in climate? As documented by Olmstead and Rhode (2011*b,a*), the answer was to plant different strains (called *cultivars*) of wheat and other grains. Where might these alternative cultivars come from? There were two main sources: Some were already being planted successfully in other regions of the world, and some were hybrid creations from existing strains of wheat, corn, etc. For example, a wheat cultivar called *Red Fife*, which originated in the Ukraine, was introduced in North America in 1842 (by David Fife and his family); and a cultivar called *Turkey*, which originated in Russia, came to North America in 1873.[6] These and other cultivars were much more resistant to the extreme temperatures of the mid-west, enabling farmers to exploit conditions that originally seemed hostile. In the U.S., the government also played an important role by testing and breeding these cultivars, and experimenting with new hybrids.[7]

Agricultural output has grown over the past two centuries in large part because of technological change. The impact of the "Green Revolution," associated with the work of people like Norman Borlaug, has been huge, but began only in the 1950s. However, technology-driven improvements in agricultural productivity began in earnest in the 1830s. As I've just explained, much of this technological change was in the development and adoption of new and hardier cultivars of grains and other plants. But it also took the form of improvements in irrigation, fertilizers, and pesticides, and the mechanization of planting and harvesting. (The book by Olmstead and Rhode (2008) documents the history of technological change in agriculture.) Technological progress in agriculture is continuing, and will be the basis for adaptation to whatever climate change occurs in the future.

7.1.3 What To Expect?

Without food, we can't live. So what will happen to agriculture as the climate changes over the coming century is of crucial importance. More than any

---

[6] These cultivars came to North America in a variety of ways. For example, as Olmstead and Rhode (2011*a*) explain, "The standard histories credit German Mennonites, who migrated from southern Russia to Kansas, with importing *Turkey* in 1873.... Furthermore, before departing for Kansas, the Mennonites tediously selected high-quality seed considered suitable for the new lands."

[7] For example, Olmstead and Rhode (2011*a*) explain how "Kansas settlers experimented with soft winter cultivars common to the eastern states, but these wheats proved unreliable in the cold winters and hot, dry summers. Tests at the Kansas Agricultural Experiment Station (AES) demonstrated *Turkey's* superiority and helped popularize the wheat. In 1919, *Turkey*-type wheat made up more than 80 percent of the wheat acreage in Nebraska and Kansas and nearly 70 percent... in Colorado and Oklahoma."

other sector of the economy, agricultural output is very sensitive to changes in the weather, and thus would seem highly vulnerable to climate change. But because climate change occurs slowly, the potential for adaptation may temper agriculture's vulnerability to it, and limit any adverse impact.

What kind and how much adaptation in agriculture should we expect in response to climate change? If climate change is severe, if summer temperatures increase substantially and winter temperatures fall, and if droughts become more severe and more frequent, will agricultural yields fall dramatically, making food much more expensive? Or will we witness adaptation to climate change that will ameliorate its impact? Will we see adaptation on the part of farmers who change what they plant and when and where they plant it; adaptation on the part of firms that might develop new hybrid strains of vegetables and fruits that are more resilient to a harsher climate; and adaptation on the part of governments that can fund research into new hybrid seeds and better methods of irrigation, and if successful, subsidize the adoption of the new technologies?

It is very likely that we will see some adaptation; after all, that is what we have seen in the past. But how much adaptation—and the overall impact of climate change on agriculture—is hard to predict. And indeed, agriculture provides an example of how hard it is to predict the impact of climate change on the economy more generally. All we can say at this point is that adaptation can help considerably to reduce the impact of climate change, so we should do what we can to promote and accelerate the technological change that will make adaptation possible.

7.2 Hurricanes, Storms, and Rising Sea Levels

Apart from its other impacts, global warming can lead to widespread flooding. This can happen via the effect of higher temperatures on the frequency and strength of hurricanes and storms, and via the effect on sea levels. As temperatures rise hurricanes and storms can become much more destructive, as warmer air provides more energy to power them. And rising temperatures can cause sea levels to rise because sea water will increase in volume as it warms, and because glaciers can melt and fragment. So should we expect to see an increase in the extent of major flooding? And how much of an increase?

The answer to the first question is yes, we should expect to see an increase in the extent of flooding. Global warming will cause at least some increase in the frequency and severity of hurricanes, and at least some increase in sea levels, both of which can cause floods. But the answer to the second question—how much of an increase in flooding—is that we don't know.

Why don't we know? First, recall from Section 5.4 that there is a great deal of uncertainty over how much sea levels might rise. It will of course depend on how much temperatures increase, but even if we could accurately predict the global mean temperature over the coming decades, there would still be considerable uncertainty as to what will happen to sea levels. Likewise, even if we could predict how much temperature will increase we still wouldn't know just how much more frequent and powerful hurricanes will become. And to start with, we can't predict how much temperatures will increase. So the bottom line is that we simply don't know the extent to which floods will become more widespread and more of a problem over the coming decades. Furthermore, the threat of flooding varies widely across regions of the world; it is less of a concern in, say, central Canada than in Bangladesh and low-lying countries in Southeast Asia.[8]

Since we don't know what will happen to sea levels, hurricanes and storms, and we therefore don't know how much of a problem flooding will become, should we sit back, relax, and wait to see what happens? Quite the contrary. As with climate change in general, it's the uncertainty itself that should push us to act now. The uncertainty creates insurance value. The fact that you don't know if or when your house may flood doesn't mean you shouldn't buy flood insurance. Flooding might become just somewhat worse or a lot worse; we don't know but it's the latter possibility we need to prepare for.

Planning for the worst is especially important when it comes to sea levels. They might rise only slightly, or they might rise a great deal, and acting now can protect us from the latter outcome. What kinds of actions can we take? Building sea walls or dikes are examples, but there are others as well, as we'll see.

7.2.1 Flooding and Its Impact

Higher sea levels and more powerful hurricanes will make flooding more likely, especially in coastal areas. How bad will things get? We don't know, so the best we can do is consider some of the possibilities, and then think about how we can protect ourselves against the worst-case scenarios.

---

[8] It is especially a problem for island nations that could end up under water, such as the Marshall Islands, Tonga, and Vanautu in the South Pacific. For an overview of the vulnerability of these and other small island nations, see Mimura (1999).

What are some of the possibilities? Several studies have made projections of the extent and global impact of rising sea levels given alternative temperature scenarios, but have arrived at a wide range of estimates.[9] This range reflects the different models that were used to come up with estimates, but it also reflects the fundamental uncertainties that we face over sea levels and over damages. However, an overall conclusion from these studies is that the potential exists for severe flooding and major damages, and we need to prepare accordingly.

What kinds of adaptation could reduce the likelihood and/or the impact of widespread flooding? We can begin by taking a cue from the substantial adaptation that has already taken place—well before concerns about climate change arose. Flooding has been a threat for a long time in many parts of the world, and adaptation to this threat occurred in a variety of ways. For example, dikes and levees (those two words mean the same thing and can be used interchangeably) have been used for many years to prevent flooding from storms and hurricanes, and to protect land that is below sea level and would otherwise be inundated with water.

An early example of such adaptation is the Dutch flood defense system, which began with the dikes that were first built on a large scale during the 13th century.[10] The materials used to build the Dutch dikes changed over time (wood was used in the early 18th century, but now the dikes consist mostly of a core of sand covered by clay to provide waterproofing and resistance against erosion), but without the dikes, much of the Netherlands would be underwater.

Likewise the New Orleans flood control system began with construction of simple levees by the French during 1717 to 1727 to prevent flooding from the Mississippi River; today the system has 192 miles of levees and 99 miles of flood walls that protect the city from the Mississippi River on one side and Lake Pontchartrain on the other. And Venice, long threatened by storms, is

[9] For example, Hinkel et al. (2014) measure flood risk in terms of expected annual damages to land and other capital and expected numbers of people flooded, and argue that Without adaptation, 0.2–4.6 percent of the global population will likely be flooded annually in 2100 under 0.25 to 1.23 meter of global mean sea level rise, with expected annual losses of 0.3–9.3 percent of global gross domestic product. Lincke and Hinkel (2018), building on Hinkel et al. (2014), find that coastal adaptation makes economic sense under five different temperature scenarios, for which sea level rise ranges from 0.3m to 2.0m. They find coastal protection is economical across all scenarios for 13 percent of the world's coastline, which hosts 90 percent of the global coastal population. And Jevrejeva et al. (2018) project that 2.0°C of warming by 2100 would cause a sea level rise of 0.63m and (with no adaptation) annual global flood damages of $11.7 trillion (13 percent of global GDP), and 3.5°C of warming would cause a sea level rise of 0.86m and annual flood damages of $14.3 trillion (16 percent of GDP). For similar studies of flooding in the U.S. see Hauer, Evans, and Mishra (2016), and for Europe see Vousdoukas et al. (2017). Desmet et al. (2021) have made global projections of flood risk by region over the next 200 years, with and without adaptation.

[10] The earliest dikes in the Netherlands appeared in the 7th century. The word *dike* was originally a term in Dutch (and other Germanic languages) that meant a long wall or embankment built to prevent flooding from the sea. For a brief history of the Dutch experience with dikes, visit http://dutchdikes.net/history/.

today protected (partially) by floodgates that are part of a flood protection system initiated in 1987. This kind of adaptation to the threat of flooding has occurred throughout the world over a long period of time. For example, although the total urban area exposed to flooding in Europe increased by around 1000 percent over the past 150 years, fatalities and economic losses (as a percentage of GDP) from flooding decreased significantly; and a similar trend of decreasing vulnerability to floods has occurred worldwide.[11]

7.2.2 Physical Barriers to Flooding

Dikes, levees, and seawalls are a form of public adaptation to the threat of flooding that continue to be widely used in many of the densely populated areas of the world. In the United States, for example, approximately 23,000 km, or 15 percent of the total coastline has been armored this way.[12] This coastline protection has been in the form of levees, and to a lesser extent, seawalls.

Unlike levees, seawalls are almost always built parallel to the shore and provide protection from wave action. In the past their primary function was erosion reduction, but now they are seen more as a defense against coastal flooding. The main advantage of a seawall is that it forms a strong coastal defense, and can thus provide a high degree of protection against coastal flooding. Seawalls also require less space than dikes and levees, especially if vertical seawall designs are used. And a seawall need not extend above the surface of the water; it can be completely submerged but still impede a storm surge.

The main disadvantage of a seawall is cost; for a given degree of protection, a seawall can be much more expensive to build. (An additional disadvantage is that sea walls can destroy shoreline habitats such as wetlands and inter-tidal beaches.) The seawall planned to encircle southern Manhattan is an example—it would prevent flooding from a storm surge like the one that occurred during Hurricane Sandy in 2012 (see Figure 1.4). $176 million in federal funding was initially allocated for the project in 2016, and later the plan was expanded to nearly $1 billion. But by 2020 the estimated cost had come to $119 billion, and the project has been shelved, at least for now.

---

[11] See Paprotny et al. (2018) for data on flooding in Europe, and Jongman et al. (2015) and Jongman (2018) for global trends.

[12] See Gittman et al. (2015). NOAA's official value for the total length of the U.S. shoreline is 95,471 miles, or about 154,000 km. This includes the shorelines of Hawaii, Alaska, offshore islands, and the Great Lakes.

The problem is that these barrier systems are expensive to build and maintain, especially if they are designed to provide protection against almost any imaginable storm or hurricane. In August 2005, Hurricane Katrina overwhelmed the New Orleans levee system and 80 percent of the city was flooded. Had the levees been built higher—at greater expense—the city might have been spared. So the problem with dikes, levees, and sea walls is how high and strong—and expensive—they should be.

In a sense, this is a problem in cost-benefit analysis—but a difficult problem. The cost of building a levee or sea wall depends on the degree of protection it is intended to provide, which can in turn depend on location, height, and engineering technology used. Given a technology and an intended degree of protection, this cost can be roughly estimated. The benefit from building a levee or sea wall is the damage (including loss of life) that it will prevent, but estimating the monetary value of that benefit is difficult. The problem is similar to that of predicting damages from climate change. As discussed in Section 3.4.2, we have little in the way of theory or data to determine the economic losses from, say, a 3°C increase in temperature. We have a bit more theory and data to draw upon when it comes to flooding, but there is still a great deal of uncertainty. No one could have predicted the arrival and strength of Hurricane Katrina in August 2005, never mind its impact on New Orleans. Furthermore, climate change means that the statistics for storm surges that might have been valid in 2005 or 2020 are unlikely to be valid a few decades from now.[13] To calculate the optimal height and technology for a particular physical barrier we need to determine the statistics for storms and storm surges that will apply in the future, which is a tall order.

Despite these difficulties, there have been efforts to develop models that could help with these kinds of cost-benefit analysis, or at least come up with rules of thumb for policy. How high should a levee be, for example? Typically a height is estimated that is expected to be breached with some specified annual probability. In the U.S., the usual policy is to choose a height for a levee that would protect against a critical water level expected to be reached or exceeded with only a 1 percent probability in any given year.

Of course if sea levels rise and hurricanes become much stronger, that critical water level will be higher than it would be otherwise, and the levee would likewise have to be higher.[14] There is evidence that there has already

---

[13] Physical barriers are built for protection in the future, and the annual probability of, say, a 10-meter storm surge in some location in 2050 is likely to be different from the probability today. See, e.g., Ceres, Forest, and Keller (2017).

[14] Ward et al. (2017) demonstrate a framework for cost-benefit analysis of river-flood risk reduction using a global flood risk model. But Ward et al. (2015) illustrate some of the limitations of these models. One of the earliest cost-benefit studies of coastal flood protection was by van Dantzig (1956).

been a significant increase in that 1 percent critical water level. For example, in August 2017, Hurricane Harvey, which caused 68 deaths and $125 billion in damages in Texas, was the state's third "500-year" flood in three years.[15]

7.2.3 Natural Barriers to Flooding

The construction of physical barriers is an important form of adaption to flood threats, but remember that nature has already provided various kinds of protection from flooding. These natural barriers include coastal wetlands that can buffer storm surges, as well as dunes and beaches, oyster and coral reefs, and maritime vegetation, all of which can break offshore waves and attenuate wave energy. But some of these natural barriers have eroded and are falling prey to property development. Reversing these trends and enhancing existing natural barriers is another way to adapt to climate change.

Nature-based solutions to increasing flood risks include widening of natural flood plains, protecting and expanding wetlands, restoring coral reefs and investing in urban green spaces to reduce run-off. A growing body of research shows that that flood protection by ecosystem creation and restoration can provide a more sustainable, cost-effective, and eco-friendly alternative to some engineering options.[16] Studies have found that hard armoring with seawalls and dikes is often detrimental to fish and wildlife habitats. For example, seawalls can disturb sea turtle nesting habitat, prevent coastal wetlands from migrating inland, and limit natural sediment buildup.

The construction of levees and dikes can be combined with the development of natural barriers. These methods can be applied mostly in places where there is a sufficient space between urbanized areas and the coastline to accommodate the creation of ecosystems that have the natural capacity to reduce storm surges.[17] Developing and enhancing natural barriers in this way can also provide added benefits such as water quality improvement, fisheries production and recreation.[18]

As Temmerman et al. (2013) show, for cities located in estuaries or deltas (such as New Orleans and London), the creation or restoration of large tidal

---

[15] For a detailed explanation of how Hurricane Harvey formed and evolved, and the damage it did, see Blake and Zelinsky (2018).

[16] See, for example, Jongman (2018), Temmerman et al. (2013), and Reguero et al. (2018). For a nice overview of nature-based solutions to flood threats, see Glick et al. (2014).

[17] See van Wesenbeeck et al. (2017) and Temmerman et al. (2013).

[18] In places like New York, New Orleans, Shanghai, Tokyo, and the Netherlands, wetlands in river deltas and estuaries were reclaimed and turned into valuable agricultural, urban, and industrial areas. However, this led to a reduction in the natural flood defenses that these wetlands had provided. See Temmerman et al. (2013), who summarize various natural adaptation strategies and where they could work.

marshes or mangroves between the city and the sea provides extra areas for water storage and can slow the movement of water. This attenuates the landward propagation of storm surges and reduces flood risks in densely populated areas. For cities behind sandy coastlines, such as Amsterdam, Abidjan in the Ivory Coast, and Lagos in Nigeria, beach and dune barriers are crucial defenses against coastal flooding. Restoring and enhancing those barriers can be an effective way to adapt to the flood threat from rising sea levels and stronger hurricanes.[19]

7.2.4 Private and Public/Private Adaptation

Adaptation to the threat of flooding can be done by households and businesses in small and not-so-small ways. At the simplest level, property developers and prospective home buyers can avoid building or buying homes in areas already vulnerable to flooding, and that will become more vulnerable as sea levels rise and hurricanes intensify. This seems like an obvious response to climate change, so why do we still see plenty of construction in these vulnerable areas? One reason is that governments subsidize insurance that makes such construction economically viable. This problem—and the need for changes in the way governments provide or subsidize flood insurance—is discussed below.

Putting aside new construction, many people face the fact that their homes are already vulnerable to the threat of flooding. What can they do to adapt to this threat? There are a variety of ways by which individuals can install systems to protect their houses from flooding. Possible strategies include the construction of a French drain (a trench containing a perforated pipe that runs along the perimeter of the house in the basement, and redirects groundwater away from the foundation), and/or one or more sump pumps to pump out any water that might have accumulated. An example of a French drain and sump pump combination is shown in Figure 7.2. Any groundwater is directed to the sump pump instead of entering the basement itself, and pumped away from the house.

There are many other ways for homes and businesses to reduce their vulnerability to flooding, include increasing the elevation of the home, using waterproof membranes and doors, "wet-flood proofing" (moving power

---

[19] Narayan et al. (2017) argues that in the U.S., natural wetlands moderated damages to New York City from Hurricane Sandy by an estimated $625 million, while the salt marshes reduced annual flood losses in Barnegat Bay in Ocean County, New Jersey, by 16 percent.

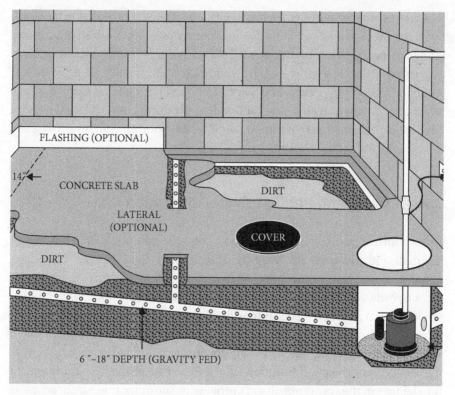

FLASHING (OPTIONAL)

14"

CONCRETE SLAB

DIRT

LATERAL
(OPTIONAL)

COVER

DIRT

6"–18" DEPTH (GRAVITY FED)

Fig. 7.2 Example of French Drain and Sump Pump. A trench containing a perforated pipe runs along the perimeter of the basement and redirects groundwater away from the foundation and toward a sump pump, where it is pumped out of the house.
Source: Arid Basement Waterproofing, https://www.aridbasementwaterproofing.com/solutions.

outlets higher on the wall and having furniture that can soak up water without getting ruined), and building flood walls. Many homeowners are unaware of these options or mistakenly think they are more expensive than they really are, so there is a role for government to provide information, improve building codes, and possibly subsidize part of the cost of retrofitting homes and buildings.[20]

Buying flood insurance would also seem like an obvious means of adaptation to the threat of flooding. At issue is the price of that insurance and who will provide it. That makes flood insurance somewhat complicated, as explained below.

---

[20] For an extensive treatment of ways to improve building codes and retrofit homes and building, see National Institute of Building Sciences (2019).

7.2.5 Flood Insurance

Homeowners can and do buy flood insurance, but for major floods, insurance is often provided at least in part by the government. In fact, because of "tail risk," i.e., the risk of an extreme event that results in huge damages, government support of some kind is usually necessary. Private insurance companies are likely to be unable or unwilling to take on this kind of catastrophic risk, and will limit the amount of insurance they will provide.[21]

What role might the government have? Governments currently provide financial compensation after large-scale disasters in a variety of ways, the simplest being direct financial support.[22] Likewise, governments at the federal or state level can establish compensation funds to (partially) indemnify victims of natural disasters (examples in the U.S. include the California Earthquake Authority and the Florida Hurricane Catastrophe Fund). Governments can also mandate coverage for floods (and other natural disasters) that is tied to other first-party insurances policies.[23] Public-private partnerships can be established and used to stimulate the availability of insurance coverage (as with the National Flood Insurance Program in the U.S.).

In the U.S., the Federal Emergency Management Agency (FEMA) manages the National Flood Insurance Program (NFIP), which is the country's main source of primary flood insurance coverage. The NFIP has more than five million policies in over 22,000 communities, collecting approximately $4.6 billion annually in premiums, fees, and surcharges. Home or business owners in a NFIP-participating community can purchase a policy if their property is located in what FEMA declares to be a high flood risk area. (Communities choose to participate in FEMA's program in order to have access to federal flood insurance, but in return they must enact minimum floodplain standards.)

At issue is what premium should NFIP (or any other government insurance program) charge? You would probably say an actuarially fair premium is appropriate, i.e., a premium equal to expected claims. Roughly, this means that the premium is equal to the probability of a claim times the amount that

---

[21] An example is government-provided flood insurance in the Netherlands. A large part of the Netherlands would be swallowed by rivers and the sea without any flood defenses. Over 60 percent of the population lives in flood-prone areas. As Jongejan and Barrieu (2008) point out, floods caused by the failure of the dikes and other defenses are high-impact, low-probability events that are hard to insure.

[22] In August 2002, for example, over a week of continuous heavy rains caused major floods in Europe, especially Germany. Compensation for flood losses was financed by the national government's disaster relief and reconstruction fund. This so-called "Sonderfonds Aufbauhilfe" amounted to 7.1 billion Euros, which was 78 percent of total direct losses.

[23] An example is the French system of tying disaster insurance to fire insurance, in place since 1982. See Magnan (1995).

will be paid in the event of a claim. Currently, insurance premiums for beach-front homes in the U.S. are far below their actuarially fair values, thereby subsidizing construction in flood-prone areas. According to a recent study by First Street Foundation (2021), the premiums would on average have to be quadrupled to reflect actuarially fair values. So why doesn't the government raise the premiums? Because that would incur the wrath of homeowners who don't want their insurance costs to rise—and who vote. It would also incur the wrath of property developers—who influence policy through their political contributions.

The problem is made even more complicated by the fact that we often can't calculate the actuarially fair premium. That calculation requires knowledge of the probabilities of claims and the corresponding amounts that would be paid, but we don't know those numbers when dealing with major "once in a hundred year" floods (which is why private insurance companies are unwilling to take on those risks). Added to that is the fact that governments are expected to step in and provide disaster relief. The result is that insurance provided in whole or in part by the government tends to subsidize construction in vulnerable areas.

In the United States, the federal government covers about 70 percent on average of the cost of recovery from major storms and floods, which is one reason why so many homes and businesses have been built in flood-prone areas where, without this subsidy, they would not have been built.[24] You might think that the NFIP and related government assistance programs were primarily designed to help the poor, but as the analysis by Ben-Shahar and Logue (2016) shows, the subsidies have been going disproportionately to wealthier households, who are more likely than the poor to own beach-front homes.

Currently economic incentives in the U.S. and elsewhere encourage building in flood-prone areas, so changing those incentives is a natural way to adapt to the storms and rising sea levels that climate change may bring about. Other actions can be taken as well. To some extent, zoning regulations and building codes either restrict land use in high risk areas or mandate certain construction rules to protect against flooding. But those regulations are often weak, and strengthening them would help to discourage building in flood-prone areas.

Changing the economic incentives along these lines seems like an obvious thing to do. But of course while it may be obvious, it is also politically difficult,

---

[24] See Gaul (2019), who estimates that Americans have chosen to build $3 trillion worth of property in some of the riskiest places in the country in terms of flooding.

given the influence of developers who benefit from the subsidies and relaxed zoning rules, and voters who don't want to see their insurance premiums go up. Then again, many aspects of climate policy are politically difficult, but are still necessary, so that one way or another the political difficulties will have to be overcome.

Finally, I have emphasized flood risk, but another impact of climate change may be droughts leading to greater risk of wildfires. An increase in the frequency and severity of wildfires has already occurred in western parts of the United States. You might think that as a result, fewer homes would be built in areas with high fire risk. But instead, more homes are being built in those areas. Why? Part of the reason is that in the U.S. construction is subsidized via public firefighting expenditures.[25] Removing the subsidy by requiring homeowners (or builders) to cover at least part of these expenditures would be one way to help us adapt to an increase in fire risk.

7.2.6 Flood Risk in Asia

For many countries, and certainly the United States, a variety of measures can be taken fairly easily to adapt to rising sea levels and stronger hurricanes, and reduce vulnerability to flooding. Dikes and levees have long been used to prevent the flooding of areas that would otherwise be underwater, and seawalls, while expensive to build and maintain, can be used to protect coastal cities from flooding. Likewise, natural barriers such as coastal wetlands, dunes, and reefs, which can provide considerable protection against storm surges, can be enhanced (and at the very least, protected). Homes and buildings can be modified to reduce their vulnerability to flooding with such things as French drains, sump pumps, and flood walls. And by pricing flood insurance so that it is actuarially fair, governments can eliminate the subsidies they currently provide for the construction of homes and buildings in areas that are likely to flood.

For some countries, however, adaptation is more difficult. This is especially the case in Asia. An example is Bangladesh, which sits on flat, low land that for the most part is just five meters above sea level. Putting aside any rise in sea levels, the country is plagued by monsoons that cause flooding almost every year, and cyclones that can arrive suddenly. In 2019, around 1.3 million homes in Bangladesh were damaged due to flooding. If sea levels rise

---

[25] Baylis and Boomhower (2019) have estimated that the present value of this subsidy can exceed 20 percent of a home's value.

significantly, flooding in Bangladesh could become catastrophic. What can be done? The country has made some progress over the past two decades by building cyclone shelters and sea walls over a limited amount of coastline. But while those adaptation efforts have been helpful, they are far from sufficient, and some have failed. An example is the construction of "polders," which are low-lying tracts of land enclosed by earthen embankments. Polders have proved to be partially effective in preventing flooding and salinity intrusion, and in protecting people and crops from the worst effects of cyclones. However, the protection they provide is limited, and without frequent and substantial maintenance, they deteriorate rapidly from erosion.

A number of large Asian coastal cities are vulnerable to flooding from rising sea levels; examples are Hong Kong and Singapore. Both cities have historically relied on developing urban drainage systems to handle large volumes of surface runoff generated during storms. This has been somewhat successful but expensive, and it is unlikely that these drainage systems as they stand could handle the impact of a one-meter or more rise in sea levels. Jakarta is another city that has been plagued by floods, largely because of uncontrolled development and a poor drainage system. (The city is actually sinking.) The Indonesian government has begun construction of a so-called "Giant Sea Wall" in an effort to prevent future flooding, but the completion date and ultimate effectiveness are unclear.

Studies of flood risk to coastal cities around the world have found that Asian port cities are most vulnerable and most likely to incur large economic losses should sea levels rise. For example, Hallegatte et al. (2013) estimate that by 2050, for moderate levels of sea level rise, 14 of the 20 cites with the highest likely economic losses from flooding (losses on the order of 1 percent of GDP annually) are Asian coastal cities.[26] Plans for adaptation, mostly via physical barriers of some kind, vary across these cities, and for some, government funding sources are limited.

7.2.7 What to Expect?

We have seen that there are some simple and inexpensive ways to adapt to flood risk, but in some cases adaptation will be difficult and expensive. Where can we expect to see considerable adaptation (and thus small potential damages from floods), and where will adaptation be more problematic?

---

[26] The cities are Guangzhou, Mumbai, Kolkata, Fukuoka-Kitakyushu, Osaka-Kobe, Shenzhen, Tianjin, Ho Chi Minh City, Jakarta, Chennai, Zhanjiang, Bangkok, Xiamen, and Nagoya. As mentioned above, Jakarta is particularly vulnerable.

The simplest and cheapest way to adapt to flood risk is to stop subsidizing construction in areas that we know are likely to flood. This shouldn't be controversial: If wealthy people want to build beach-front vacation homes in places where they are likely to be washed away by the next hurricane, that's fine, but do we really think it makes sense for taxpayers to help pay for those homes? (Developers and wealthy homeowners would probably say yes, but hopefully that's a minority view that won't drive climate policy.)

The next step is more expensive, but important. We need to repair and enhance existing systems of levees, as was recently done with the levees surrounding New Orleans. (This can be thought of as part of infrastructure investment.) And we need to plan and begin construction of sea walls around vulnerable cities, which in the U.S. range from large ones like New York and Houston to smaller ones like Charleston, South Carolina and Norfolk, Virginia. And for some cities, existing sea walls need to be enhanced (for example, the Thames Barrier, that currently protects London from flooding, would fail if sea levels rise substantially). Where possible, we also need to preserve and enhance natural barriers to coastal flooding.

But in some parts of the world adaptation will be more difficult, and the next steps to take are less clear. The Asian coastal cities discussed above are an example of this problem. An even more extreme example is the set of small island countries that could end up completely underwater should sea levels rise sufficiently. For those island countries, adaptation (beyond migration of the entire population) may simply not be possible.

7.3 Solar Geoengineering

We have looked at a variety of ways to reduce the impact of potential or actual flooding, which can be brought about by rising sea levels and stronger and more frequent hurricanes. Sea walls, dikes, and related measures deal with some of the harmful impacts of climate change. Another form of adaptation is quite different: *solar geoengineering*, which would reduce the warming effects of any buildup of atmospheric CO_2. Solar geoengineering can be done in different ways, but the approach that is viewed as most promising is quite simple: "Seed" the atmosphere, at an altitude of roughly 20 kilometers, or 70,000 feet, with sulfur or sulfur dioxide. These "seeds" would remain in the atmosphere for up to a year, after which time they would precipitate as sulfuric acid and fall back to earth. (Thus the "seeding" would have to be repeated regularly.) While in the atmosphere the particles would reduce the greenhouse effect by reflecting sunlight back into space.

Remember that all of the CO_2 in the atmosphere will remain there. The only thing the sulfur dioxide will do is cause the atmosphere to reflect more sunlight, which will nullify some of the harmful warming effects of the CO_2. In technical terms, the sulfur dioxide increases the *albedo*, i.e., reflectivity, of the Earth's atmosphere.[27]

Solar geoengineering might seem expensive, but it's not. Yes, the sulfur dioxide will eventually come down from the atmosphere in the form of sulfuric acid, so that the "seeding" will have to be repeated every year or even more frequently. But the cost of the seeding itself is low. And that low cost creates another advantage: It partly eliminates the free-rider problem that makes emissions abatement so difficult. Rather than reducing its own emissions (at considerable cost), a country like India can "free ride" on the emission reductions of other countries. And because it is so cheap, solar geoengineering doesn't require the participation of all, or even most countries—it could be done effectively by just a few countries.

This approach to solar geoengineering, which uses sulfur or a sulfur compound for atmospheric seeding, is usually referred to as *Stratospheric Aerosol Injection* (SAI). But other methods have also been proposed, using other aerosols, and in different ways. (See Figure 7.3.) For example, *Marine Cloud Brightening* would use a sea salt or related compound to seed low-lying clouds over the ocean, with the goal of increasing the reflectivity of the clouds. Another approach, *Cirrus Cloud Thinning*, would reduce the density of high-altitude cirrus clouds by seeding them with an aerosol, thereby allowing more heat to escape back into space. But while these other methods are interesting, SAI is currently viewed as by far the most feasible, so I will focus on it here.[28] In their 2014 and 2018 reports, the IPCC views solar geoengineering as promising, but notes that the costs (which I discuss below) are currently uncertain. (The IPCC uses the term "Solar Radiation Management" rather than solar geoengineering.)

Solar geoengineering is rarely viewed as a cure-all for the problem of climate change. First, it has some potential problems. Although it would reduce the warming effects of the build-up of CO_2 in the atmosphere, it would not reduce the build-up itself, so that the CO_2 concentration would keep increasing. That's a problem because (as discussed later) there is a serious

---

[27] Albedo is measured on a scale of 0 to 1, where 0 means no reflectivity (all light is absorbed) and 1 means complete reflectivity (no light absorbed). The average albedo of the Earth's atmosphere has been estimated to be about 0.30.

[28] Still other proposed methods include space-based reflectors, tropospheric aerosols, and increasing the reflectivity of crops or other land cover. However, these other approaches are currently viewed as far less feasible. See, e.g., Kravitz and MacMartin (2020) and National Academies of Sciences and Medicine (2021).

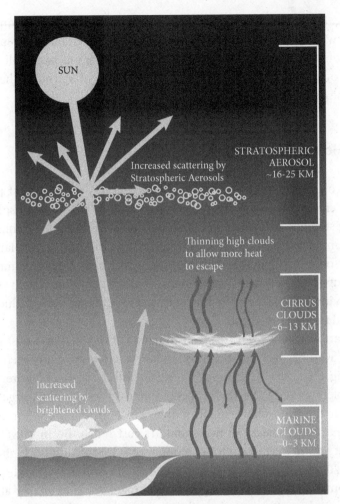

Fig. 7.3 Alternative Approaches to Solar Geoengineering. Most promising is *Stratospheric Aerosol Injection* (SAI), which uses sulfur or a sulfur compound for atmospheric seeding at an altitude of about 20 km. More speculative approaches include *Marine Cloud Brightening*, which would seed low-lying clouds to increase their reflectivity; and *Cirrus Cloud Thinning*, which would reduce the density of high-altitude cirrus clouds by seeding them with an aerosol.
Source: National Academies of Sciences and Medicine (2021), p. 32.

risk that the additional CO_2 would cause acidification of the world's oceans. Also, unless accompanied by a reduction in CO_2 emissions, the atmospheric seeding would have to continue indefinitely, because once it stopped, the global mean temperature would rise.

Given that the seeding would have to go on indefinitely, geoengineering is largely viewed as a temporary and partial solution to the problem of global warming. It might be used to reduce the temperature increase by one or two degrees, with the idea that it could eventually be phased out and replaced by sharp reductions in net CO_2 emissions. Thus in what follows, I will examine the use of the technology to reduce the amount of warming that would otherwise occur by 1°C.

7.3.1 How It Would Work

The approach that has been most well-studied and appears most practical involves creating a cloud of sulfuric acid (H_2SO_4) in the stratosphere. By blocking sunlight, this cloud would reduce radiative forcing and thus reduce the amount of warming due to atmospheric CO_2. This effect is similar to what happens following large volcanic eruptions, which spew huge amounts of SO_2 into the upper atmosphere. Indeed, major volcanic eruptions (such as Mount Pinatubo in 1991) have resulted in significant (but temporary) reductions in the global mean temperature.[29]

There are alternative ways to create the cloud of H_2SO_4, but the use of sulfur dioxide (SO_2) has received the most attention and is currently seen as most promising. Once it is in the atmosphere, SO_2 will combine with water to form tiny droplets of H_2SO_4. So, this process seems easy—just inject a large amount of SO_2 into the atmosphere, and let it do its job. Furthermore sulfur (and thus SO_2) is extremely cheap, so this shouldn't cost much. But first we need to figure out how to inject a sufficient amount of SO_2 at an altitude high enough (around 20 km) for the material to circulate in the atmosphere and stay there for 6 months to a year. You might think the answer is to simply allocate a bunch of airplanes for the task, and if there aren't enough airplanes available to do the job, just build some more.

The problem is that airplanes currently in use were designed to fly at altitudes no greater than 40,000 feet, which is about 12 km. That means new airplanes must be designed and built that are capable of flying at a much higher altitude. How much would it cost and how long would it take before such airplanes become available? The cost estimates vary widely, but it is generally acknowledged that it would take at least 10 years, and maybe 15, before any airplanes become available.[30]

---

[29] For a detailed discussion of the connection between geoengineering and volcanic eruptions, see Robock (2000).

[30] To convert kilometers to miles, 1 kilometer is 0.621 miles, or 3,281 feet. So 20 km is equal to 65,620 feet. Some have proposed the use of balloons instead of airplanes to deliver the SO_2; see, e.g., Davidson et al. (2012). But the current view is the balloons would be far less efficient and not cost effective. The advantages of using airplanes to deliver the SO_2 are discussed in Robock et al. (2009).

Suppose the airplanes to deliver the SO_2 were available now. And suppose further that the objective is not to eliminate all warming, but rather to reduce the increase in the global mean temperature that would otherwise occur by 1°C (so that a 3°C increase in temperature would be reduced to only a 2°C increase). In that case, how many airplanes would we need, and how would we use them?

Estimates by Smith and Wagner (2018), Smith, Dykema, and Keith (2018), and Keith, Wagner, and Zabel (2017), along with summaries in Intergovernmental Panel on Climate Change (2018), suggest that counteracting a 1°C temperature increase would require maintaining around 10 Mt (million tons) of SO_2 in the stratosphere (which would react with water vapor to form an aerosol cloud of about 15 Mt of H_2SO_4). And how would we inject 10 Mt of SO_2 into the stratosphere? One way to do this is to burn (within the airplane engines) about 5 Mt of molten sulfur and then release the resulting 10 Mt of SO_2 at the required 20 km altitude.[31] The SO_2 and H_2SO_4 would eventually dissipate, so the injections of SO_2 would have to be repeated every 6 to 12 months in order to maintain the aerosol cloud.

Repeatedly injecting 10 Mt of SO_2 into the stratosphere would require a large number of new airplanes—depending on their size, as many as 300— that would operate almost nonstop. But the good news is that we could build up to the 10 Mt level very gradually. The reason is that increases in the atmospheric CO_2 concentration, and hence increases in temperature, will occur very slowly, over a period of decades. Our concern is preventing an increase in the global mean temperature of 3°C by the end of the century—not by the year 2030 or 2040.

Allowing for 10 to 15 years to design, test, and build the first airplanes, actual injections of SO_2 might start around 2035, and could ramp up slowly, perhaps over the following 20 or 30 years. Smith and Wagner (2018) outline a modest scenario in which six new airplanes, together capable of carrying a payload of 0.1 Mt of sulfur (and release 0.2 Mt of SO_2), are built and put into service each year. After 10 years, the fleet of 60 planes would be sufficient to deliver 2 Mt of SO_2 per year. Eventually we would need 300 planes to deliver 10 Mt of SO_2 per year, but the ramping up could occur slowly. This is consistent with the idea that solar geoengineering is not something we would do now to fix an immediate problem. Rather, it is an option that we would exercise in the future if we saw the CO_2 concentration and the temperature rising faster than we anticipated.

[31] By mass, 15 Mt of H_2SO_4 = 10 Mt SO_2 = 5 Mt of sulfur, S.

7.3.2 How Much Would It Cost?

Again, suppose we already had a sufficient number of airplanes capable of flying at 20 km. If that were the case, the cost of injecting 10 Mt of SO_2 into the stratosphere would be very low—essentially just the cost of operating the airplanes plus the cost of the sulfur and other materials. But we don't have the required airplanes, and that's where the real cost—and uncertainty over the cost—come in.

A few studies claim that existing aircraft could inject SO_2 at the required altitude after some modifications, making the total cost very low. But most other studies assume that the airplanes capable of doing the job would have to be designed and then built.[32] How much would it cost to design the planes, and then build them? And how long would it take? The last question is easiest to answer: It would take around 10 to 15 years to design and build the required number of planes. As for the cost, all we have right now is a series of rough estimates, and they differ considerably.

Given the uncertainty over the cost of developing and building the airplanes, should we conclude that solar geoengineering might turn out to be very expensive? No, quite the contrary. Even if we take the highest estimates for the cost of the planes—and then double that cost—the annual total cost of using solar geoengineering to reduce the temperature increase that would otherwise occur by 1°C would be very low. How low? Estimates of the total annual cost, including the amortized cost of the airplanes, range from $20 billion (Smith and Wagner (2018)) to $40 billion (de Vries, Janssens, and Hulshoff (2020)) to a high of about $110 billion (Robock et al. (2009)).[33] Thus it would be reasonable to put the cost in the (admittedly wide) range of $20 to $200 billion per year.

Let's take the upper end of that range, $200 billion per year. That might seem like a great deal of money, but put in the context of the climate problem we face, and alternative ways of dealing with that problem, it's not. Remember, the objective here is to eliminate a 1°C increase in the *global* mean temperature, so the $200 billion must be compared to global GDP, which in 2020 was about

---

[32] McClellan, Keith, and Apt (2012) argues that with modifications, existing aircraft could do the job. But Smith and Wagner (2018) review a range of lofting methods and conclude that a new plane design is needed because no existing aircraft have the combination of altitude and payload capabilities required. Their paper relies in part on personal communications that the authors had with 13 commercial aerospace vendors, and assumes that the new airplanes could be developed and produced in 15 years.

[33] Robock (2020) summarizes these cost estimates, and compares them by assuming that in each case the cost of developing and building the airplanes would be amortized over 20 years, uses the payload cost estimates from de Vries, Janssens, and Hulshoff (2020). Cost estimates in Keith, Wagner, and Zabel (2017) are also consistent with this range.

$90 trillion. That means that the $200 billion annual cost amounts to only about 0.2 percent of GDP, a tiny fraction.[34] To put this in context, compare it to the cost of using a carbon tax to prevent the global mean temperature from rising more than 2°C. There is uncertainty about this, but we would probably need a global carbon tax on the order of $100 per ton to reduce CO_2 emissions sufficiently to meet a 2°C target. With global CO_2 emissions of about 37 Gt in 2020, that carbon tax would add up to nearly $4 trillion, which is close to 4 percent of global GDP. That's some 20 times greater than the cost of solar geoengineering.

Put simply, solar geoengineering could be a very cost-effective way to prevent increases in the atmospheric concentration of CO_2 from substantially increasing the global mean temperature. Even assuming a $200 billion annual cost, which is about double the highest recent estimate, this is a very small fraction of the cost of any alternative (such as a broadly based carbon tax). And the actual cost might turn out to be closer to $20 billion per year rather than $200 billion. The low cost of solar geoengineering does not mean that it is the answer to climate change; it has some serious problems, as I discuss below. But in terms of feasibility and cost effectiveness, it should certainly be part of our climate toolbox.

7.3.3 Problems with Solar Geoengineering

Solar geoengineering is controversial, to put it mildly. Many environmentalists view it as anathema, but usually for the wrong reason. Their concern is that solar geoengineering, as well as other forms of adaptation, will deflect us from doing what we should be doing to reduce GHG emissions. After all, if we know that there is a cheaper and easier alternative, why go to the effort and considerable expense of reducing emissions? There is some truth to this concern, but remember that I have not been arguing that we should give up on reducing emissions. I am arguing instead that solar geoengineering should be something that we consider and be prepared to use *in addition* to reducing emissions.

There are other concerns about solar geoengineering. The main concern it that it might create its own set of environmental problems. The most

[34] For comparison, in 1998 the U.S. Energy Information Administration estimated that the cost of complying with the Kyoto Procol—which was intended to keep the increase in the global mean temperature to 3°C—would be about 2 percent of GDP. (See Energy Information Administration (1998).) Estimates of country cost studies assembled by Intergovernmental Panel on Climate Change (2007, 2014) also put the cost at around 2 percent of GDP. Furthermore, the use of solar geoengineering would also limit the uncertainty we have over future temperature increases, which itself is valuable, as explained in Pindyck (2014).

important of these potential problems is that if we use it to prevent the temperature from increasing, CO_2 will continue to accumulate in the atmosphere, and some of it will be absorbed by the oceans, making them more acidic. I will turn to this issue of ocean acidification below, but first, here are some of the other environmental concerns that have been raised:

- **Impact on Rainfall.** There is concern that solar geoengineering might result in a reduction in global mean precipitation, and/or may affect precipitation patterns in some regions of the world. Some of this concern is based on the decrease in precipitation that was observed after the eruption of Mount Pinatubo in 1991. Climate models are mixed in what they predict about precipitation, but the impact is likely to be limited if solar engineering is used only to prevent part of the temperature increase, e.g., the 1°C discussed above.[35]

- **Vegetation and Crop Yields.** There might be an impact on vegetation, both through changes in the hydrologic cycle and feedbacks on plant physiology. Multiple studies focus on how vegetation might respond to a high-CO_2, low temperature climate in contrast to the current low-CO_2, low temperature climate. Studies to date, however, generally point in the same direction: solar geoengineering can improve global crop yields.[36]

- **Ozone Depletion and Health Impacts.** There is a concern that Stratospheric Aerosol Injection could lead to further ozone depletion via changes it would bring about in the chemistry of the stratosphere. See, for example, Tilmes et al. (2009) and Weisenstein, Keith, and Dykema (2015). Also, materials injected into the stratospheric could have health impacts when washed down if they enter food and water supplies. Effiong and Neitzel (2016) reviews the medical literature on possible health impacts of various SRM aerosols.

- **Governance.** The fact that solar geoengineering is so cheap is a big plus, but it also creates a problem. Who decides whether to pursue it, and to what degree? Because it is cheap, a small group of countries, or even one country (e.g., the U.S.), could do it alone, without the kind of international agreement needed to reduce global CO_2 emissions. Do we need a set of rules, in the form of a treaty or other international

---

[35] See, for example, Kleidon, Kravitz, and Renner (2015). Irvine et al. (2019) examine an SG scenario that "halves the warming from doubling CO_2, and roughly restores the intensity of the hydrological cycle, rather than the typical scenario in which SG offsets all warming." In that case, their model shows neither temperature, water availability, extreme temperature, nor extreme precipitation are exacerbated.

[36] See, for example, Pongratz et al. (2012), Cao (2018), Dagon and Schrag (2019), and Tjiputra, Grini, and Lee (2016).

agreement, specifying when and how solar geoengineering could be used? Because the impact of global warming can vary so much across countries, reaching such an agreement is likely to be difficult. (See the papers in Harvard Project on Climate Agreements (2019) for a discussion of these issues.)

- **The Stopping Problem.** Suppose we use solar geoengineering to prevent a 1°C increase in the global mean temperature, and we do this by injecting around 10 Mt of SO_2 into the stratosphere (which would react with water vapor to form an aerosol cloud of about 15 Mt of H_2SO_4). Remember that we would have to maintain that aerosol cloud by continually injecting more SO_2 as the H_2SO_4 gradually dissipates. What if we stop replenishing the aerosol cloud? The atmospheric CO_2 concentration would remain high, so the result would be a rapid increase in temperature. Without some kind of international commitment to maintain the SO_2 injections, this "stopping problem" can be a serious risk. The problem is illustrated in Figure 7.4, which shows simulations of eight different climate models of average surface temperature and average precipitation when solar radiation management (SRM) is used to counter the warming effects of a 1 percent per year increase in the CO_2 concentration for 50 years (solid lines), and without solar geoengineering (dashed lines). Note that when SRM is stopped, temperature and precipitation increase rapidly to where they would be without SRM.

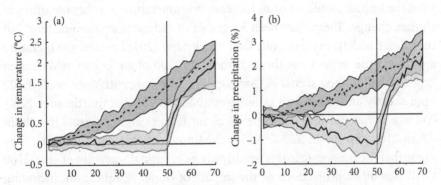

Fig. 7.4 The Stopping Problem. Eight models were used to simulate the change in globally averaged (a) surface temperature and (b) precipitation. Solid lines are for simulations using Solar Radiation Management (SRM) to balance a 1 percent per year increase in CO_2 concentration until year 50, after which SRM is stopped. Dashed lines are for simulations with a 1 percent per year increase in CO_2 concentration and no SRM. The shaded areas show the 25th to 75th percentiles from the models.
Source: Intergovernmental Panel on Climate Change (2014), *The Physical Science Basis*, page 634.

Ocean Acidification

Apart from the problems listed above, there is one other major problem, namely ocean acidification. While at this point the process is poorly understood, we know that as the atmospheric CO_2 concentration increases, some of the CO_2 is absorbed into the oceans, which can reduce their average pH (i.e., "acidify" the oceans). Because solar geoengineering would do nothing to limit the increase in the atmospheric CO_2 concentration, it would do nothing to prevent ocean acidification. That is probably the strongest argument for why reducing CO_2 emissions is likely to be a better policy instrument than geoengineering.[37]

But there are uncertainties here, of two types. First, to what extent would increases in the atmospheric CO_2 concentration cause changes in the average pH of the oceans? Second, what would be the economic and ecological impact of any reductions in the pH that might occur?

Several studies have used Earth system models to try to project the effect of increases in the atmospheric CO_2 concentration through the end of the century on ocean pH. For example, Bopp et al. (2013) summarize the results of 10 Earth system models, and show that by the end of the century, the average pH could fall by as much as 0.30, from a current average value of about 8.1.[38] Even if global CO_2 emissions are reduced substantially, the average pH could fall by at least 0.1 unit.

Suppose the average pH drops by 0.30 units by the end of the century. What would be the impact of that drop? We don't know, just as we don't know what the impact would be of an increase in temperature or other measures of climate change. There have been a number of estimates and projections, but they differ widely. For example, Colt and Knapp (2016) review the literature and assess the impact out through the year 2200 of an "ocean acidification catastrophe." In their scenario, "Atmospheric CO_2 concentrations reach 1,000 ppm shortly after 2100 and stabilize at about 2,000 ppm shortly after 2200. Average ocean pH levels decline by about 0.3 per century, reaching about 7.8 in 2100 and 7.5 in 2200." As for the impact of that decline in pH, they argue that the economic losses would only be around 0.1 percent of year-2100 GDP. But other projections of the impact of ocean acidification, including those of Intergovernmental Panel on Climate Change (2014), are much more pessimistic.

---

[37] In addition, the sulfuric acid that eventually rains down from the stratosphere can make lakes and rivers more acidic. But this has not been a major concern with solar geoengineering.

[38] Also see the discussion in Williamson and Turley (2012). A pH of 7.0 is neutral, so 8.1 is slightly basic (or alkaline). The pH was about 8.2 before the industrial revolution, so has already fallen by 0.10 units.

7.3.4 What to Do?

As I explained at the outset of this book, of all the policy options we could consider, solar geoengineering is by far the most controversial. Many environmentalists consider it to be downright dangerous, and dismiss it out of hand. This is partly due to the concern about ocean acidification, but more generally a concern that doing anything to alter the environment is risky and must be avoided. For some environmentalists, emitting CO_2 and other GHGs into the atmosphere must be avoided, and seeding the atmosphere with a sulfur compound must likewise be avoided.

On top of their fears about possible negative impacts of solar geoengineering (and other forms of adaptation, such as sea walls), some environmentalists will argue that once we decide we can adapt to climate change, society won't want to spend resources on costly measures to reduce emissions. This is a valid point, and would be even more valid if we knew that we can reduce emissions fast enough and far enough to prevent higher temperature and rising sea levels. But we don't know that we can reduce emissions fast enough and far enough. On the contrary, as I have argued throughout this book, it is very unlikely that the world will do what is needed to prevent the global mean temperature from rising more than 2°C. Not impossible, but very unlikely. And that means that we need to do what we can to prepare for the possibility of a very bad climate outcome.

So what should we do with solar geoengineering? There are certainly concerns about its use, as outlined above. But at the same time, it could well turn out to be a cheap and effective tool that we should have at the ready if our efforts to sharply reduce global CO_2 emissions turn out badly. Solar geoengineering cannot be done today, because we do not yet have the capability of seeding the atmosphere with sulfur or sulfur compounds at a sufficiently high altitude. And the atmospheric CO_2 concentration is not yet at a level where we need solar geoengineering. But a consensus is building that we should undertake much more research in this area, and do so quickly. This would include R&D directed at alternative sulfur and non-sulfur based aerosols, and possible impacts on rainfall, ozone depletion, and ocean acidity. In addition, we should start now to develop and build the airplanes that will be needed should be have to turn to this tool in the future.

7.4 Can Adaptation Solve Our Climate Problem?

No, of course not! We simply don't know the extent to which adaptation can help us reduce our vulnerability to climate change, but we can be fairly sure that while it will help, it won't eliminate the problem. As we saw in the case of rising sea levels, we can stop subsidizing construction in high flood-risk areas, strengthen or build levees and sea walls, and build or retrofit homes to protect against flooding. But those steps won't provide complete protection. If sea levels rise by one or two meters and hurricanes become much stronger on average than they are today, we will see flooding, especially in regions that are close to (or even below) sea level. And the damage from flooding could be especially acute for countries like Bangladesh, Thailand, Vietnam, or some of small island nations that may end up under water. As for solar geoengineering, we have seen that it carries some risks (most notably with respect to ocean acidification), so more research is needed.

On the other hand, we do know that adaptation can help in a variety of important ways. As we have seen, its impact on agriculture has been profound; it has enabled crop yields to steadily increase, despite (in the U.S.) the climate "change" that was part of the movement west. And levees, dikes, and sea walls have already been successfully used to protect large areas from inundation—the Netherlands, much of which would be underwater without its dikes, is a prime example. And we have every reason to think that despite the risks, solar geoengineering can help, and in fact may be necessary at least temporarily, if temperatures rise more than we now anticipate.

Won't the very possibility of adaptation distract us from the important task of reducing GHG emissions as quickly and extensively as possible? It need not, as long as we treat adaptation as just part of a comprehensive climate policy. And even if it does somewhat lower the political pressure to rapidly reduce emissions, we would be derelict to ignore the importance of adaptation or even delay taking steps to facilitate its implementation. We simply can't put ourselves in a position where, despite our best efforts, the global mean temperature rises more than 2°C, the impact is extreme, and we are unable to respond.

Some adaptation will happen by itself, especially adaptation by households and private firms. But some, such as sea walls and solar geoengineering, will require government action at the federal, state, and local levels. And this will take time, so we need to begin soon to work on the planning and R&D that are needed to make adaptation successful.

7.5 Climate Future

At the risk of being unduly repetitive, I will stress once more that I am not claiming in this book that reducing global emissions sufficiently to prevent a temperature increase of more than 2°C is impossible. We don't know the extent to which various countries will reduce their emissions, as part of or apart from some kind of international agreement that might eventually be reached. We might indeed reach an international agreement and see large reductions in emissions, but that is not something we should count on. And even if we could be sure that emissions will be reduced sharply and quickly, we don't know what result it will have for temperature change, sea levels, and other aspects of climate. The question is whether society should take the risk of being unprepared should things will turn out badly. I have argued that doing so would be a big mistake.

Where does this leave us? Here is a quick summary of the arguments that I've made in this book:

(1) Yes, we should do all we can to reduce net GHG emissions, hopefully as efficiently (i.e., cost effectively) as possible. We can do this with a carbon tax or cap-and-trade system, with government subsidies and mandates, and by extracting CO_2 from the atmosphere or from power plant exhausts.

(2) But "we" means the world. If the U.S., U.K., and Europe reduce emissions but other countries do not, we'll be in bad shape, and the outlook won't be good. We must reduce *global* emissions, which will probably require an international agreement, adherence to which is observable and enforceable.

(3) Even with such an international agreement, we cannot count on preventing a temperature increase well above 2°C. We need to face the reality that the atmospheric CO_2 concentration is likely to grow for many years, pushing temperatures up. Maybe we'll be lucky and the temperature won't increase that much, but maybe we'll be very unlucky and witness a temperature increase of 3° or even 4°C.

(4) And if the temperature does increase by 3°C or more, what will that warming do to the economy? And what will be the effect of warming on sea levels and the strength of hurricanes? Most important, what will be the overall impact of higher temperatures and rising sea levels on GDP and other measures of human welfare over the coming decades? We don't know. Maybe we'll be lucky and the impact will be small, but maybe we'll be very unlucky and the impact will be extreme.

(5) We need to prepare for the possibility that we will be very unlucky, and that we will find ourselves heading towards a climate catastrophe. In that case we will need to rely more than we had anticipated on adaptation. But some kinds of adaptation require more research (as

with solar geoengineering), and will take time for planning and the early stages of implementation (sea walls, solar geoengineering). That means we need to invest in R&D now, and take other necessary steps to make adaptation timely and effective.

Climate (and Other) Catastrophes

Remember that our concern has been the risk of a climate catastrophe—a very bad outcome, not simply in terms of higher temperatures, but in terms of its impact on the economy and human welfare generally. But once we start thinking about catastrophes, we need to go beyond climate and recognize that we face a variety of other kinds of potential catastrophes that could occur, and do great damage to society and human welfare. What kinds of catastrophes? At the risk of ending this book on a depressing note, here are a few examples:

- Major pandemics. Have you heard about COVID-19? The Spanish Flu? The CDC promises that more pandemics are likely to arrive. We've learned a lot from COVID-19, including how to develop vaccines at amazing speed. That should help when the next pandemic arrives, but that next one might be far more virulent.
- Bioterrorism. Have you heard of anthrax? Sarin? Might terrorists get their hands on new biological or chemical agents? We don't know, but a bioterrorist attack could result in many deaths and a panic-induced shock to GDP.
- Nuclear terrorism. What would be the impact? Possibly a million or more deaths, and a huge shock to GDP from a reduction in trade and economic activity worldwide, as vast resources are devoted to averting further events.
- Nuclear war. Countries that already have nuclear weapons are building up their stockpiles, and more countries are likely to acquire a nuclear capability in the future. Enough nuclear weapons already exist to wipe out every human on the planet several times over. Will some of those weapons actually be used? Good question.
- Cyber warfare that would cause major damage to our energy and financial systems, and to our infrastructure more generally. We have already seen some (fortunately limited) examples of this; imagine it occurring on a grand scale.
- Other catastrophic risks. You can use your imagination here. The following events are less likely but certainly catastrophic: gamma ray bursts, an asteroid hitting the Earth, and unforeseen consequences of AI or nanotechnology.

Climate change receives a lot of our attention, as well it should. It is often called an existential threat to humanity, and it may indeed be. After all, as I have stressed throughout this book, a catastrophic climate outcome is a real possibility. But there are other existential threats that are also real possibilities, and also deserve our attention. These other threats get less attention, and we tend to ignore them. One or more of the potential catastrophes listed above could occur sooner and have a greater impact than climate change, and yet we are not doing as much as we could to prevent them or prepare for the possibility of their occurrence.[39]

Why don't we do more to avert these other potential catastrophes? And why—despite all the attention it receives—don't we do more to avert a climate catastrophe? I believe the answer, at least in part, is that as individuals and as a society we are inherently myopic. Or put another way, we discount far-future costs and benefits at a very high rate. Climate change might do serious damage, but not this year, and probably not for another couple of decades. For most of us, that's just very far off, so we prefer not to think about it. And for politicians, who have an aversion to unpleasant things like taxes, it's so far off that it can be ignored.

The risk of a climate (or other) catastrophe is inherently a long-run problem, and dealing with it requires a long-run outlook. It won't be easy, but we—the public and the politicians—will have to overcome our natural myopia, and direct much more of our attention to the decades that lie ahead.

7.6 Further Readings

Throughout this book I have stressed the fact that despite our best efforts, CO_2 emissions are unlikely to fall fast enough to prevent an increase in the global mean temperature that is greater, perhaps much greater, than the widely cited 2°C limit. This means that we may face rising sea levels, more frequent and stronger hurricanes and storms, and other adverse climate effects. We must prepare for the possibility of that outcome by investing in adaptation. Adaptation comes in many forms; in this chapter I focused on the development and adoption of heat- and drought-resistant crops, measures such as the construction of levees and sea walls to reduce the possible impact of flooding from rising sea levels and stronger hurricanes, and the use of solar

---

[39] The argument that we need to devote more resources to the prevention of these other potential catastrophes was made by Posner (2004), among others. And which of these potential catastrophes should get the highest priority? For an analysis of that question, see Martin and Pindyck (2015, 2021).

geoengineering to reduce the warming effect of a rising atmospheric CO_2 concentration. The discussion has been brief, and there is much more to read.

- I explained that when it comes to agriculture, adaptation to climate change has been happening steadily over nearly 200 years, as people moved west and had to grow crops in climates that had more extreme temperature and rainfall fluctuations. Adaptation occurred through the development and introduction of new cultivars, and new productivity-enhancing technologies for planting and harvesting crops. The book by Olmstead and Rhode (2008) documents the history of technological change in agriculture, and Mellor (2017) discusses the development and impact of agriculture in low- and middle-income countries. What kinds of innovations in agriculture can we expect over the coming decades? Read the book by Lee (2019) for some speculative but extremely interesting answers to that question.

- There is considerable uncertainty over the extent to which sea levels will rise, but we need to prepare for that possibility, along with the risks of flooding from stronger and more frequent hurricanes. One way to do this is through the construction of physical barriers such as sea walls and levees. But we should also protect and enhance natural barriers to flooding such as coastal wetlands, dunes, coral reefs, and maritime vegetation. For an overview of nature-based solutions to flood threats, see Glick et al. (2014). And for an interesting explanation of how Hurricane Harvey formed and devastated parts of Texas, see Blake and Zelinsky (2018).

- I argued that government insurance programs subsidize the construction of homes and businesses in flood-prone (and wildfire-prone) areas, because the premiums that are charged are below those that would be actuarially fair. The book by Gaul (2019) and the report by First Street Foundation (2021) provide thorough and readable treatments of this problem. Also, the First Street Foundation website (https://firststreet.org) provides an online tool to calculate flood risk for any town or county in the United States, and shows the discrepancy between insurance premiums and actuarial risks.

- I briefly discussed ways that building codes can be changed and homes and buildings can be retrofitted to reduce their vulnerability to flooding. For a deep dive into retrofit strategies and ways to improve building codes, see National Institute of Building Sciences (2019).

- Barrett (2008) provides one of the earliest introductions to solar geoengineering and why it is an important policy tool. For overviews of the technology and how it would work, including some of the issues that need to be resolved, see Irvine et al. (2016), Smith, Dykema, and Keith (2018), Keith and Irvine (2019), and Robock (2020). For a nice summary paper on the likely cost of solar geoengineering, see Smith and Wagner (2018). For general discussions of the promises, problems, and outlook for solar geoengineering, see the conference volume, Harvard Project on Climate Agreements (2019) and the recent book, National Academies of Sciences and Medicine (2021).

• The recent National Climate Assessment provides an overview of the effects of climate change that we have already experienced in the U.S. For the published version, see U.S. Global Change Research Program (2018).

• Climate change has already occurred over the past thousands of years, including the extensive warming that followed the last ice age. In addition to adapting to those changes, humans have had to adapt to large regional differences in climate as they migrated across the world. Examining the historical record on adaptation can help us predict the extent to which humans will be able to adapt to *future* climate change. For an interesting account of how farmers in Europe adapted to the warming that occurred over the past several thousand years, see Fagan (2008), and for an overview of the measurement of adaptation based on the historical record, see Massetti and Mendelsohn (2018).

• I concluded by explaining that there are other potential catastrophes that we need to worry about, and do more to prepare for. The books by Posner (2004) and Bostrom and Ćirković (2008) discuss a variety of potential catastrophes, and are guaranteed to keep you up at night. For a detailed (and frightening) treatment of nuclear terrorism, see Allison (2004).

Bibliography

Aldrin, Magne, Marit Holden, Peter Guttorp, Ragnhild Bieltvedt Skeie, Gunnar Myhre, and Terje Koren Berntsen. 2012. "Bayesian Estimation of Climate Sensitivity Based on a Simple Climate Model Fitted to Observations of Hemispheric Temperatures and Global Ocean Heat Content." *Environmetrics*, 23.

Aldy, Joseph E., Alan J. Krupnick, Richard G. Newell, Ian W. H. Parry, and William A. Pizer. 2010. "Designing Climate Mitigation Policy." *Journal of Economic Literature*, 48(4): 903–934.

Aldy, Joseph E., and Richard J. Zeckhauser. 2020. "Three Prongs for Prudent Climate Policy." National Bureau of Economic Research Working Paper 26991.

Allen, Myles R., and David J. Frame. 2007. "Call Off the Quest." *Science*, 318: 582–583.

Allen, Myles R., Jan S. Fuglestvedt, Keith P. Shine, Andy Reisinger, Raymond T. Pierrehumbert, and Piers M. Forster. 2016. "New Use of Global Warming Potentials to Compare Cumulative and Short-Lived Climate Pollutants." *Nature Climate Change*, 6(8): 773–776.

Allison, Graham. 2004. *Nuclear Terrorism: The Ultimate Preventable Catastrophe.* Henry Holt & Company.

Alvarez, Ramón A., Daniel Zavala-Araiza, David R. Lyon, David T. Allen, Zachary R. Barkley, Adam R. Brandt, et al. 2018. "Assessment of Methane Emissions from the U.S. Oil and Gas Supply Chain." *Science*, 361(6398): 186–188.

Amazon Fund. 2010. "Amazon Fund Activity Report 2010." Brazilian National Development Bank Report.

Amazon Fund. 2019. "Amazon Fund Activity Report 2019." Brazilian National Development Bank Report.

Andrews, Timothy, Jonathan M. Gregory, David Paynter, Levi G. Silvers, Chen Zhou, Thorsten Mauritsen, Mark J. Webb, Kyle C. Armour, Piers M. Forster, and Holly Titchner. 2018. "Accounting for Changing Temperature Patterns Increases Historical Estimates of Climate Sensitivity." *Geophysical Research Letters*, 45(16): 8490–8499.

Annan, J. D., and J. C. Hargreaves. 2006. "Using Multiple Observationally-Based Constraints to Estimate Climate Sensitivity." *Geophysical Research Letters*, 33(6): 1–4.

Arrow, Kenneth J., and Anthony C. Fisher. 1974. "Environmental Preservation, Uncertainty, and Irreversibility." *The Quarterly Journal of Economics*, 88(2): 312–319.

Auffhammer, Maximilian. 2018. "Quantifying Economic Damages from Climate Change." *Journal of Economic Perspectives*, 32(4): 33–52.

Auffhammer, Maximilian, Solomon M. Hsiang, Wolfram Schlenker, and Adam Sobel. 2013. "Using Weather Data and Climate Model Output in Economic Analyses of Climate Change." *Review of Environmental Economics and Policy*, 7: 181–198.

Baccini, A., S. J. Goetz, W. S. Walker, N. T. Laporte, Mindy Sun, Damien Sulla-Menashe, Joe Hackler, P. S. A. Beck, Ralph Dubayah, M. A. Friedl, et al. 2012. "Estimated Carbon Dioxide Emissions from Tropical Deforestation Improved by Carbon-Density Maps." *Nature Climate Change*, 2(3): 182–185.

Barrett, Scott. 2008. "The Incredible Economics of Geoengineering." *Environmental and Resource Economics*, 39: 45–54.

Bastin, Jean-Francois, Yelena Finegold, Claude Garcia, Danilo Mollicone, Marcelo Rezende, Devin Routh, Constantin M. Zohner, and Thomas W. Crowther. 2019. "The Global Tree Restoration Potential." *Science*, 365(6448): 76–79.

Baylis, Patrick, and Judson Boomhower. 2019. "Moral Hazard, Wildfires, and the Economic Incidence of Natural Disasters." National Bureau of Economic Research Working Paper 26550.

Ben-Shahar, Omri, and Kyle D. Logue. 2016. "The Perverse Effects of Subsidized Weather Insurance." *Stanford Law Review*, 68: 571–626.

Blake, Eric S., and David A. Zelinsky. 2018. "Hurricane Harvey." National Hurricane Center Tropical Cyclone Report AL092017.

Blanc, Elodie, and Wolfram Schlenker. 2017. "The Use of Panel Models in Assessments of Climate Impacts on Agriculture." *Review of Environmental Economics and Policy*, 11(2): 258–279.

Blue Ribbon Commission on America's Nuclear Future. 2012. "Report to the Secretary of Energy." Blue Ribbon Commission Technical Report.

Bopp, Laurent, Laure Resplandy, James C. Orr, Scott C. Doney, John P. Dunne, M. Gehlen, P. Halloran, Christoph Heinze, Tatiana Ilyina, Roland Seferian, et al. 2013. "Multiple Stressors of Ocean Ecosystems in the 21st Century: Projections with CMIP5 Models." *Biogeosciences*, 10(10): 6225–6245.

Bostrom, Nick, and Milan Ćirković, ed. 2008. *Global Catastrophic Risks*. Oxford University Press.

Brown, Patrick T., and Ken Caldeira. 2017. "Greater Future Global Warming Inferred from Earth's Recent Energy Budget." *Nature*, 552(7683): 45.

Burke, Marshall, and Kyle Emerick. 2016. "Adaptation to Climate Change: Evidence from U.S. Agriculture." *American Economic Journal: Economic Policy*, 8(3): 106–140.

Burke, Marshall, John Dykema, David B. Lobell, Edward Miguel, and Shanker Satyanath. 2015. "Incorporating Climate Uncertainty into Estimates of Climate Change Impacts." *Review of Economics and Statistics*, 97(2): 461–471.

Cain, Michelle, John Lynch, Myles R. Allen, Jan S. Fuglestvedt, David J. Frame, and Adrian H. Macey. 2019. "Improved Calculation of Warming-Equivalent Emissions for Short-Lived Climate Pollutants." *Climate and Atmospheric Sciences*, 2(29): 1–7.

Cai, Yongyang, and Thomas S. Lontzek. 2019. "The Social Cost of Carbon with Economic and Climate Risks." *Journal of Political Economy*, 127(6): 2684–2734.

Cao, Long. 2018. "The Eeffects of Solar Radiation Management on the Carbon Cycle." *Current Climate Change Reports*, 4(1): 41–50.

Ceres, Robert L., Chris E. Forest, and Klaus Keller. 2017. "Understanding the Detectability of Potential Changes to the 100-Year Peak Storm Surge." *Climatic Change*, 145(1): 221–235.

Chen, Cuicui, and Richard Zeckhauser. 2018. "Collective Action in an Asymmetric World." *Journal of Public Economics*, 158: 103–112.

Cline, William R. 2020. "Transient Climate Response to Cumulative Emissions (TCRE) As A Reduced-Form Climate Model." Economics International, Inc. Working Paper 20-02.

Coady, David, Ian Parry, Nghia-Piotr Le, and Baoping Shang. 2019. "Global Fossil Fuel Subsidies Remain Large: An Update Based on Country-Level Estimates." International Monetary Fund Working Paper 19/89.

Colt, Stephen G, and Gunnar P. Knapp. 2016. "Economic Effects of an Ocean Acidification Catastrophe." *American Economic Review*, 106(5): 615–619.

Cox, Peter M., Chris Huntingford, and Mark S. Williamson. 2018. "Emergent Constraint on Equilibrium Climate Sensitivity from Global Temperature Variability." *Nature*, 553(7688): 319.

Crowther, Thomas W., Henry B. Glick, Kristofer R. Covey, Charlie Bettigole, Daniel S. Maynard, Stephen M. Thomas, Jeffrey R. Smith, Gregor Hintler, Marlyse C. Duguid, Giuseppe Amatulli, et al. 2015. "Mapping Tree Density at a Global Scale." *Nature*, 525(7568): 201–205.

Dagon, Katherine, and Daniel P. Schrag. 2019. "Quantifying the Effects of Solar Geoengineering on Vegetation." *Climatic Change*, 153(1–2): 235–251.

d'Annunzio, Rémi, Marieke Sandker, Yelena Finegold, and Zhang Min. 2015. "Projecting Global Forest Area Towards 2030." *Forest Ecology and Management*, 352: 124–133.

Davidson, Peter, Chris Burgoyne, Hugh Hunt, and Matt Causier. 2012. "Lifting Options for Stratospheric Aerosol Geoengineering: Advantages of Tethered Balloon Systems." *Philosophical Transactions of the Royal Society A: Mathematical, Physical and Engineering Sciences*, 370(1974): 4263–4300.

Dell, Melissa, Benjamin F. Jones, and Benjamin A. Olken. 2012. "Temperature Shocks and Economic Growth: Evidence from the Last Half Century." *American Economic Journal: Macroeconomics*, 4: 66–95.

Dell, Melissa, Benjamin F. Jones, and Benjamin A. Olken. 2014. "What Do We Learn from the Weather? The New Climate-Economy Literature." *Journal of Economic Literature*, 52(3): 740–798.

Deschênes, Olivier, and Michael Greenstone. 2007. "The Economic Impact of Climate Change: Evidence from Agricultural Output and Random Fluctuations in Weather." *American Economic Review*, 97(1): 1–15.

Deschênes, Olivier, and Michael Greenstone. 2011. "Climate Change, Mortality, and Adaptation: Evidence from Annual Fluctuations in Weather in the US." *American Economic Journal: Applied Economics*, 3: 152–185.

Desmet, Klaus, Robert E. Kopp, Scott A. Kulp, Dávid Krisztián Nagy, Michael Oppenheimer, Esteban Rossi-Hansberg, and Benjamin H. Strauss. 2021. "Evaluating the Economic Cost of Coastal Flooding." *American Economic Journal: Macroeconomics*.

Dessler, A. E., and P. M. Forster. 2018. "An Estimate of Equilibrium Climate Sensitivity from Interannual Variability." *Journal of Geophysical Research: Atmospheres*, 123(16): 8634–8645.

de Vries, Iris E., M. Janssens, and S. J. Hulshoff. 2020. "A Specialised Delivery System for Stratospheric Sulphate Aerosols (Part 2): Financial Cost and Equivalent CO_2 Emissions." *Climatic Change*, 162(1): 87–103.

Diaz, Delavane, and Frances Moore. 2017. "Quantifying the Economic Risks of Climate Change." *Nature Climate Change*, 7: 774–782.

Dixit, Avinash K., and Robert S. Pindyck. 1994. *Investment Under Uncertainty*. Princeton University Press.

Effiong, Utibe, and Richard L. Neitzel. 2016. "Assessing the Direct Occupational and Public Health Impacts of Solar Radiation Management with Stratospheric Aerosols." *Environmental Health*, 15(1): 7.

Emanuel, Kerry. 2018. *What We Know about Climate Change*. Cambridge, MA: MIT Press.

Energy Information Administration. 1998. "Comparing Cost Estimates for the Kyoto Protocol." U.S. Government Report 09-45.

Fagan, Brian. 2008. *The Great Warming: Climate Change and the Rise and Fall of Civilizations*. New York: Bloomsbury Press.

First Street Foundation. 2021. "The Cost of Climate: America's Growing Flood Risk." First Street Foundation Technical Report.

Fisher, Anthony C., W. Michael Hannemann, Michael J. Roberts, and Wolfram Schlenker. 2012. "The Economic Impacts of Climate Change: Evidence from Agricultural Output and Random Fluctuations in Weather: Comment." *American Economic Review*, 102(7): 3749–3760.

Food and Agriculture Organization. 2020. "State of the World's Forests." United Nations Report.

Franklin, Sergio L., Jr., and Robert S. Pindyck. 2018. "Tropical Forests, Tipping Points, and the Social Cost of Deforestation." *Ecological Economics*, 153: 161–171.

Frederick, Shane. 2006. "Valuing Future Life and Future Lives: A Framework for Understanding Discounting." *Journal of Economic Psychology*, 27: 667–680.

Freeman, Mark C., Gernot Wagner, and Richard Zeckhauser. 2015. "Climate Sensitivity Uncertainty: When Is Good News Bad?" *Philosophical Transactions A*, 373: 1–15.

Fuss, Sabine. 2017. "The 1.5 C Target, Political Implications, and the Role of BECCS." In *Oxford Research Encyclopedia of Climate Science*.

Fuss, Sabine, Chris D. Jones, Florian Kraxner, Glen Philip Peters, Pete Smith, Massimo Tavoni, Detlef Peter van Vuuren, Josep G. Canadell, Robert B. Jackson, J. Milne, et al. 2016. "Research Priorities for Negative Eemissions." *Environmental Research Letters*, 11(11): 115007.

Gates, Bill. 2021. *How to Avoid a Climate Disaster.* New York: Knopf.

Gaul, Gilbert M. 2019. *The Geography of Risk: Epic Storms, Rising Seas, and the Cost of America's Coasts.* Sarah Crichton Books.

Gibbs, Holly K., Sandra Brown, John O. Niles, and Jonathan A. Foley. 2007. "Monitoring and Estimating Tropical Forest Carbon Stocks: Making REDD a Reality." *Environmental Research Letters*, 2(4): 045023.

Gillett, Nathan P., Vivek K. Arora, Damon Matthews, and Myles R. Allen. 2013. "Constraining the Ratio of Global Warming to Cumulative CO2 Emissions Using CMIP5 Simulations." *Journal of Climate*, 26(18): 6844–6858.

Gillingham, Kenneth, William Nordhaus, David Anthoff, Geoffrey Glanford, Valentina Bosetti, Peter Christensen, Haewon McJeon, and John Riley. 2018. "Modeling Uncertainty in Integrated Assessment of Climate Change: A Multimodel Comparison." *Journal of the Association of Environmental and Resource Economists*, 5(4): 791–826.

Gittman, Rachel K., F. Joel Fodrie, Alyssa M. Popowich, Danielle A. Keller, John F. Bruno, Carolyn A. Currin, Charles H. Peterson, and Michael F. Piehler. 2015. "Engineering Away our Natural Defenses: An Analysis of Shoreline Hardening in the U.S." *Frontiers in Ecology and the Environment*, 13(6): 301–307.

Glick, Patty, John Kostyack, James Pittman, Tania Briceno, and Nora Wahlund. 2014. "Natural Defenses from Hurricanes and Floods: Protecting America's Communities and Ecosystems in an Era of Extreme Weather." National Wildlife Federation Report.

Gollier, Christian. 2001. *The Economics of Risk and Time.* Cambridge, MA: M.I.T. Press.

Gollier, Christian. 2013. *Pricing the Planet's Future.* Princeton University Press.

Gollier, Christian. 2019. *Le Climat après la Fin du Mois.* Paris, France: Presses Universitaires de France.

Greenstone, Michael, Elizabeth Kopits, and Ann Wolverton. 2013. "Developing a Social Cost of Carbon for Use in U.S. Regulatory Analysis: A Methodology and Interpretation." *Review of Environmental Economics and Policy*, 7(1): 23–46.

Hallegatte, Stephane, Colin Green, Robert J. Nicholls, and Jan Corfee-Morlot. 2013. "Future Flood Losses in Major Coastal Cities." *Nature Climate Change*, 3(9): 802–806.

Hansen, James, Makiko Sato, Paul Hearty, Reto Ruedy, Maxwell Kelley, Valerie Masson-Delmotte, Gary Russell, George Tselioudis, Junji Cao, Eric Rignot, et al. 2016. "Ice Melt, Sea Level Rise and Superstorms: Evidence from Paleoclimate Data, Climate Modeling, and Modern Observations that 2°C Global Warming Could Be Dangerous." *Atmospheric Chemistry and Physics*, 16(6): 3761–3812.

Harris, Nancy L., Sandra Brown, Stephen C. Hagen, Sassan S. Saatchi, Silvia Petrova, William Salas, Matthew C. Hansen, Peter V. Potapov, and Alexander Lotsch. 2012. "Baseline Map of Carbon Emissions from Deforestation in Tropical Regions." *Science*, 336(6088): 1573–1576.

Harvard Project on Climate Agreements. 2019. "Governance of the Deployment of Solar Geoengineering." Harvard University Report.

Hassler, John, Per Krusell, and Conny Olovsson. 2018. "The Consequences of Uncertainty: Climate Sensitivity and Economic Sensitivity to the Climate." *Annual Review of Economics*, 10: 189–205.

Hauer, Mathew E., Jason M. Evans, and Deepak R. Mishra. 2016. "Millions Projected to Be at Risk from Sea-Level Rise in the Continental United States." *Nature Climate Change*, 6(7): 691–695.

Hawkins, Ed, and Rowan Sutton. 2009. "The Potential to Narrow Uncertainty in Regional Climate Predictions." *Bulletin of the American Meterological Society*, 90: 1095–1107.

Heal, Geoffrey. 2017a. "The Economics of the Climate." *Journal of Economic Literature*, 55(3): 1–18.

Heal, Geoffrey. 2017b. "What Would It Take to Reduce U.S. Greenhouse Gas Emissions 80 Percent by 2050?" *Review of Environmental Economics and Policy*, 11(2): 319–335.

Heal, Geoffrey. 2020. "Economic Aspects of the Energy Transition." National Bureau of Economic Research Working Paper 27766.

Heal, Geoffrey, and Antony Millner. 2014. "Uncertainty and Decision Making in Climate Change Economics." *Review of Environmental Economics and Policy*, 8(1): 120–137.

Health Effects Institute. 2020. "State of Global Air 2020." Health Effects Institute Special Report.

Hegerl, Gabriele C., Thomas J. Crowley, William T. Hyde, and David J. Frame. 2006. "Climate Sensitivity Constrained by Temperature Reconstructions over the Past Seven Centuries." *Nature*, 440: 1029–1032.

Hepburn, Cameron, Ella Adlen, John Beddington, Emily A. Carter, Sabine Fuss, Niall Mac Dowell, Jan C. Minx, Pete Smith, and Charlotte K. Williams. 2019. "The Technological and Economic Prospects for CO_2 Utilization and Removal." *Nature*, 575(7781): 87–97.

High-Level Commission on Carbon Prices. 2017. *Report of the High-Level Commission on Carbon Prices*. Washington, D.C. : World Bank.

Hinkel, Jochen, Daniel Lincke, Athanasios T. Vafeidis, Mahé Perrette, Robert James Nicholls, Richard S. J. Tol, Ben Marzeion, Xavier Fettweis, Cezar Ionescu, and Anders Levermann. 2014. "Coastal Flood Damage and Adaptation Costs under 21st Century Sea-Level Rise." *Proceedings of the National Academy of Sciences*, 111(9): 3292–3297.

Holland, Stephen P., Erin T. Mansur, and Andrew J. Yates. 2020. "The Electric Vehicle Transition and the Economics of Banning Gasoline Vehicles." National Bureau of Economic Research Working Paper 26804.

Holland, Stephen P., Jonathan E. Hughes, Christopher R. Knittel, and Nathan C. Parker. 2015. "Some Inconvenient Truths About Climate Change Policy: The Distributional Impacts of Transportation Policies." *Review of Economics and Statistics*, 97(5): 1052–1069.

Hope, C. W. 2006. "The Marginal Impact of CO2 from PAGE 2002: An Integrated Assessment Model Incorporating the IPCC's Five Reasons for Concern." *The Integrated Assessment Journal*, 6: 19–56.

Houghton, John. 2015. *Global Warming: The Complete Briefing, 5th Ed.* Cambridge University Press.

Houghton, R. A., Brett Byers, and Alexander A. Nassikas. 2015. "A Role for Tropical Forests in Stabilizing Atmospheric CO_2." *Nature Climate Change*, 5(12): 1022–1023.

Hsiang, Solomon, and Robert E. Kopp. 2018. "An Economist's Guide to Climate Change Science." *Journal of Economic Perspectives*, 32(4): 3–32.

Interagency Working Group on Social Cost of Carbon. 2013. "Technical Support Document: Technical Update of the Social Cost of Carbon for Regulatory Impact Analysis." U.S. Government.

Intergovernmental Panel on Climate Change. 2007. *Climate Change 2007*. Cambridge University Press.

Intergovernmental Panel on Climate Change. 2014. *Climate Change 2014*. Cambridge University Press.

Intergovernmental Panel on Climate Change. 2018. *Global Warming of 1.5°C (Special Report)*. World Meteorological Organization and United Nations Environment Program.

Intergovernmental Panel on Climate Change. 2021. *Climate Change 2021: The Physical Science Basis*. Cambridge University Press.

International Energy Agency. 2019. "Nuclear Power in a Clean Energy System." International Energy Agency Technical Report.

Irvine, Peter J., Ben Kravitz, Mark G. Lawrence, and Helene Muri. 2016. "An Overview of the Earth System Science of Solar Geoengineering." *Wiley Interdisciplinary Reviews: Climate Change*, 7(6): 815–833.

Irvine, Peter, Kerry Emanuel, Jie He, Larry W. Horowitz, Gabriel Vecchi, and David Keith. 2019. "Halving Warming with Idealized Solar Geoengineering Moderates Key Climate Hazards." *Nature Climate Change*, 9(4): 295–299.

Jacobsen, Mark, Christopher R. Knittel, James Sallee, and Arthur van Benthem. 2020. "Sufficient Statistics for Imperfect Externality-Correcting Policies." *Journal of Political Economy*, 128(5): 1826–1876.

Jarvis, Stephen, Olivier Deschenes, and Akshaya Jha. 2019. "The Private and External Costs of Germany's Nuclear Phase-Out." National Bureau of Economic Research Working Paper 26598.

Jevrejeva, Svetlana, L. P. Jackson, Aslak Grinsted, Daniel Lincke, and Ben Marzeion. 2018. "Flood Damage Costs under the Sea Level Rise with Warming of 1.5 C and 2 C." *Environmental Research Letters*, 13(7): 074014.

Jongejan, Ruben, and Pauline Barrieu. 2008. "Insuring Large-Scale Floods in the Netherlands." *The Geneva Papers on Risk and Insurance—Issues and Practice*, 33(2): 250–268.

Jongman, Brenden. 2018. "Effective Adaptation to Rising Flood Risk." *Nature Communications*, 9(1): 1–3.

Jongman, Brenden, Hessel C. Winsemius, Jeroen C. J. H. Aerts, Erin Coughlan De Perez, Maarten K. Van Aalst, Wolfgang Kron, and Philip J. Ward. 2015. "Declining Vulnerability to River Floods and the Global Benefits of Adaptation." *Proceedings of the National Academy of Sciences*, 112(18): E2271–E2280.

Keery, John S., Philip B. Holden, and Neil R. Edwards. 2018. "Sensitivity of the Eocene Climate to CO_2 and Orbital Variability." *Climate of the Past*, 14(2): 215–238.

Keith, David, and Peter Irvine. 2019. "The Science and Technology of Solar Geoengineering: A Compact Summary." In *Governance of the Deployment of Solar Geoengineering*. Cambridge, MA: Harvard Project on Climate Agreements.

Keith, David W., and John M. Deutch. 2020. "Climate Policy Enters Four Dimensions." In *Securing Our Economic Future*, ed. Melissa S. Kearney and Amy Ganz. The Aspen Institute.

Keith, David W., Geoffrey Holmes, David St Angelo, and Kenton Heidel. 2018. "A Process for Capturing CO_2 from the Atmosphere." *Joule*, 2(8): 1573–1594.

Keith, David W., Gernot Wagner, and Claire L. Zabel. 2017. "Solar Geoengineering Reduces Atmospheric Carbon Burden." *Nature Climate Change*, 7(9): 617–619.

Keohane, Nathaniel O. 2009. "Cap and Trade Rehabilitated: Using Tradeable Permits to Control U.S. Greenhouse Gases." *Review of Environmental Economics and Policy*, 3(1): 42–62.

King, Mervyn. 2016. *The End of Alchemy: Money, Banking, and the Future of the Global Economy*. New York: W. W. Norton & Company.

Kleidon, Axel, Ben Kravitz, and Maik Renner. 2015. "The Hydrological Sensitivity to Global Warming and Solar Geoengineering Derived from Thermodynamic Constraints." *Geophysical Research Letters*, 42(1): 138–144.

Knoblauch, Christian, Christian Beer, Susanne Liebner, Mikhail Grigoriev, and Eva-Maria Pfeiffer. 2018. "Methane Production as Key to the Greenhouse Gas Budget of Thawing Permafrost." *Nature Climate Change*, 8: 309–312.

Knutti, Reto, Maria A. A. Rugenstein, and Gabriele C. Hegerl. 2017. "Beyond Equilibrium Climate Sensitivity." *Nature Geoscience*, 10: 727–744.

Kolstad, Charles D. 1996. "Fundamental Irreversibilities in Stock Externalities." *Journal of Public Economics*, 60: 221–233.

Kolstad, Charles D. 2010. *Environmental Economics, 2nd Edition*. New York: Oxford University Press.

Kolstad, Charles D., and Frances C. Moore. 2020. "Estimating the Economic Impacts of Climate Change Using Weather Observations." *Review of Environmental Economics and Policy*, 14(1): 1–24.

Koonin, Steven E. 2021. *Unsettled: What Climate Science Tells Us, What It Doesn't, and Why It Matters*. Dallas: BenBella Books.

Kopits, Elizabeth, Alex Marten, and Ann Wolverton. 2013. "Incorporating 'Catastrophic' Climate Change into Policy Analysis." *Climate Policy*, 14(5): 637–664.

Kopp, Robert E., Radley M. Horton, Christopher M. Little, Jerry X. Mitrovica, Michael Oppenheimer, D. J. Rasmussen, Benjamin H. Strauss, and Claudia Tebaldi. 2014. "Probabilistic 21st and 22nd Century Sea-Level Projections at a Global Network of Tide-Gauge Sites." *Earth's Future*, 2(8): 383–406.

Kotchen, Matthew J. 2018. "Which Social Cost of Carbon? A Theoretical Perspective." *Journal of the Association of Environmental and Resource Economists*, 5(3): 673–694.

Kotchen, Matthew J. 2021. "The Producer Benefits of Implicit Fossil Fuel Subsidies in the United States." *Proceedings of the National Academy of Sciences*, 118(14).

Kravitz, Ben, and Douglas G. MacMartin. 2020. "Uncertainty and the Basis for Confidence in Solar Geoengineering Research." *Nature Reviews Earth & Environment*, 1(1): 64–75.

Krekel, Daniel, Remzi Can Samsun, Ralf Peters, and Detlef Stolten. 2018. "The Separation of CO_2 from Ambient Air—A Techno-Economic Assessment." *Applied Energy*, 218: 361–381.

Krissansen-Totton, Joshua, and David C. Catling. 2017. "Constraining Climate Sensitivity and Continental versus Seafloor Weathering Using an Inverse Geological Carbon Cycle Model." *Nature Communications*, 8: 15423.

Larson, E., C. Greig, J. Jenkins, E. Mayfield, A. Pascale, C. Zhang, J. Drossman, R. Williams, S. Pacala, R. Socolow, E. J. Baik, R. Birdsey, R. Duke, R. Jones, B. Haley, E. Leslie, K. Paustian, and A. Swan. 2020. "Net-Zero America: Potential Pathways, Infrastructure, and Impacts." Princeton University Interim Report.

Lee, Shen Ming. 2019. *Hungry For Disruption: How Tech Innovations Will Nourish 10 Billion By 2050*. New Degree Press.

Le Quéré, Corinne, Robert B. Jackson, Matthew W. Jones, Adam J. P. Smith, Sam Abernethy, Robbie M. Andrew, Anthony J. De-Gol, David R. Willis, Yuli Shan, Josep G. Canadell, et al. 2020. "Temporary Reduction in Daily Global CO_2 Emissions During the COVID-19 Forced Confinement." *Nature Climate Change*, 10(7): 647–653.

Lewis, Nicholas, and Judith Curry. 2018. "The Impact of Recent Forcing and Ocean Heat Uptake Data on Estimates of Climate Sensitivity." *Journal of Climate*, 31(15): 6051–6071.

Libardoni, Alex G., and Chris E. Forest. 2013. "Correction to 'Sensitivity of Distributions of Climate System Properties to the Surface Temperature Data Set'." *Geophysical Research Letters*, 40(10): 2309–2311.

Li, Canbing, Haiqing Shi, Yijia Cao, Yonghong Kuang, Yongjun Zhang, Dan Gao, and Liang Sun. 2015. "Modeling and Optimal Operation of Carbon Capture from the Air Driven by Intermittent and Volatile Wind Power." *Energy*, 87: 201–211.

Lincke, Daniel, and Jochen Hinkel. 2018. "Economically Robust Protection against 21st Century Sea-Level Rise." *Global Environmental Change*, 51: 67–73.

Litterman, Bob. 2013. "What Is the Right Price for Carbon Emissions?" *Regulation*, 36(2): 38–43.

Lohmann, Ulrike, and David Neubauer. 2018. "The Importance of Mixed-Phase and Ice Clouds for Climate Sensitivity in the Global Aerosol-Climate Model ECHAM6-HAM2." *Atmospheric Chemistry and Physics*, 18(12): 8807–8828.

Lomborg, Bjorn. 2020. *False Alarm: How Climate Change Panic Costs Us Trillions, Hurts the Poor, and Fails to Fix the Planet.* New York: Basic Books.

Lynch, John, Michelle Cain, Raymond Pierrehumbert, and Myles Allen. 2020. "Demonstrating GWP*: A Means of Reporting Warming-Equivalent Emissions that Captures the Contrasting Impacts of Short- and Long-Lived Climate Pollutants." *Environmental Research Letters,* 15(044023).

Magnan, Serge. 1995. "Catastrophe Insurance System in France." *Geneva Papers on Risk and Insurance: Issues and Practice,* 474–480.

Manski, Charles F. 2013. *Public Policy in an Uncertain World: Analysis and Decisions.* Harvard University Press.

Manski, Charles F. 2020. "The Lure of Incredible Certitude." *Economics & Philosophy,* 36(2): 216–245.

Markandya, Anil, and Paul Wilkinson. 2007. "Electricity Generation and Health." *The Lancet,* 370(9591): 979–990.

Martin, Ian W. R., and Robert S. Pindyck. 2015. "Averting Catastrophes: The Strange Economics of Scylla and Charybdis." *American Economic Review,* 105(10): 2947–2985.

Martin, Ian W. R., and Robert S. Pindyck. 2021. "Welfare Costs of Catastrophes: Lost Consumption and Lost Lives." *The Economic Journal,* 131(634): 946–969.

Massetti, Emanuele, and Robert Mendelsohn. 2018. "Measuring Climate Adaptation: Methods and Evidence." *Review of Environmental Economics and Policy,* 12(2): 324–341.

Matthews, H. Damon, Kirsten Zickfeld, Reto Knutti, and Myles R. Allen. 2018. "Focus on Cumulative Emissions, Global Carbon Budgets and the Implications for Climate Mitigation Targets." *Environmental Research Letters,* 13(1).

Matthews, H. Damon, Nathan P. Gillett, Peter A. Scott, and Kirsten Zickfeld. 2009. "The Proportionality of Global Warming to Cumulative Carbon Emissions." *Nature,* 459(11): 829–833.

McClellan, Justin, David W. Keith, and Jay Apt. 2012. "Cost Analysis of Stratospheric Albedo Modification Delivery Systems." *Environmental Research Letters,* 7(3): 034019.

Mellor, John W. 2017. *Agricultural Development and Economic Transformation: Promoting Growth with Poverty Reduction.* Palgrave Studies in Agricultural Economics and Food Policy.

Mendelsohn, Robert, William D. Nordhaus, and Daigee Shaw. 1994. "The Impact of Global Warming on Agriculture: A Ricardian Analysis." *American Economic Review,* 84(4): 753–771.

Mengel, Matthias, Anders Levermann, Katja Frieler, Alexander Robinson, Ben Marzeion, and Ricarda Winkelmann. 2016. "Future Sea Level Rise Constrained by Observations and Long-Term Commitment." *Proceedings of the National Academy of Sciences,* 113(10): 2597–2602.

Metcalf, Gilbert E. 2009. "Market-Based Policy Options to Control U.S. Greenhouse Gas Emissions." *Journal of Economic Perspectives,* 23(2).

Metcalf, Gilbert E. 2019. *Paying for Pollution: Why a Carbon Tax is Good for America.* New York: Oxford University Press.

Mimura, Nobuo. 1999. "Vulnerability of Island Countries in the South Pacific to Sea Level Rise and Climate Change." *Climate Research,* 12(2–3): 137–143.

Morgan, M. Granger, Parth Vaishnav, Hadi Dowlatabadi, and Ines L. Azevedo. 2017. "Rethinking the Social Cost of Carbon Dioxide." *Issues in Science and Technology,* 43–50.

Müller, Ulrich K., James H. Stock, and Mark W. Watson. 2019. "An Econometric Model of International Long-Run Growth Dynamics." National Bureau of Economic Research Working Paper 26593.

Mullins, Jamie T., and Prashant Bharadwaj. 2021. "Weather, Climate, and Migration in the United States." National Bureau of Economic Research Working Paper 28614.

Narayan, Siddharth, Michael W. Beck, Paul Wilson, Christopher J. Thomas, Alexandra Guerrero, Christine C. Shepard, Borja G. Reguero, Guillermo Franco, Jane Carter Ingram, and Dania Trespalacios. 2017. "The Value of Coastal Wetlands for Flood Damage Reduction in the Northeastern USA." *Scientific Reports*, 7(1): 1–12.

National Academies of Sciences, Engineering, and Medicine. 2021. *Reflecting SunLight: Recommendations for Solar Geoengineering Research and Research Governance*. Washington, D.C.: The National Academies Press.

National Academy of Sciences. 2017. *Valuing Climate Damages: Updating Estimation of the Social Cost of Carbon Dioxide*. Washington, D.C.: National Academies Press.

National Institute of Building Sciences. 2019. "National Hazard Mitigation Saves: 2019 Report." National Institute of Building Sciences Technical Report.

National Research Council. 2015. *Climate Intervention: Carbon Dioxide Removal and Reliable Sequestration*. Washington, D.C.: National Academies Press.

Nordhaus, William D. 1991. "To Slow or Not to Slow: The Economics of the Greenhouse Effect." *Economic Journal*, 101: 920–937.

Nordhaus, William D. 1993. "Optimal Greenhouse Gas Reductions and Tax Policy in the 'DICE' Model." *American Economic Review*, 83: 313–317.

Nordhaus, William D. 2008. *A Question of Balance: Weighing the Options on Global Warming Policies*. Yale University Press.

Nordhaus, William D. 2013. *The Climate Casino*. Yale University Press.

Nordhaus, William D. 2015. "Climate Clubs: Overcoming Free-Riding in International Climate Policy." *American Economic Review*, 105(4): 1339–1370.

Nordhaus, William D. 2018. "Projections and Uncertainties about Climate Change in an Era of Minimal Climate Policies." *American Economic Journal: Economic Policy*, 10(3): 333–360.

Nordhaus, William D. 2019. "Climate Change: The Ultimate Challenge for Economics." *American Economic Review*, 109(6): 1991–2014.

Nordhaus, William D, and Andrew Moffat. 2017. "A Survey of Global Impacts of Climate Change: Replication, Survey Methods, and a Statistical Analysis." National Bureau of Economic Research Working Paper 23646.

Occidental Petroleum Corporation. 2020. "Pathway to Net-Zero." Occidental Climate Report 2020.

Olmstead, Alan L., and Paul W. Rhode. 2008. *Creating Abundance: Biological Innovation and American Agricultural Development*. Cambridge University Press.

Olmstead, Alan L., and Paul W. Rhode. 2011a. "Adapting North American Wheat Production to Climatic Challenges, 1839–2009." *Proceedings of the National Academy of Sciences*, 108(2): 480–485.

Olmstead, Alan L., and Paul W. Rhode. 2011b. "Responding to Climatic Challenges: Lessons from U.S. Agricultural Development." In *The Economics of Climate Change: Adaptations Past and Present*. 169–94. University of Chicago Press.

Olsen, Roman, Ryan Sriver, Marlos Goes, Nathan Urban, H. Damon Matthews, Murali Haran, and Klaus Keller. 2012. "A Climate Sensitivity Estimate Using Bayesian Fusion of Instrumental Observations and an Earth System Model." *Journal of Geophysical Research: Atmospheres*, 117(4).

Paltsev, Sergey, Andrei Sokolov, Henry Chen, Xiang Gao, Adam Schlosser, Erwan Monier, Charles Fant, Jeffery Scott, Qudsia Ejaz, Evan Couzo, et al. 2016. "Scenarios of Global Change: Integrated Assessment of Climate Impacts." MIT Joint Program on Global Change Report 291.

Paprotny, Dominik, Antonia Sebastian, Oswaldo Morales-Nápoles, and Sebastiaan N. Jonkman. 2018. "Trends in Flood Losses in Europe over the Past 150 Years." *Nature Communications*, 9(1): 1985.

Peters, G. P., S. J. Davis, and R. Andrew. 2012. "A Synthesis of Carbon in International Trade." *Biogeosciences*, 9: 3247–3276.

Phaneuf, Daniel J., and Till Requate. 2017. *A Course in Environmental Economics: Theory, Policy, and Practice.* Cambridge, U.K.: Cambridge University Press.

Pindyck, Robert S. 1993. "Investments of Uncertain Cost." *Journal of Financial Economics*, 34(1): 53–76.

Pindyck, Robert S. 2000. "Irreversibilities and the Timing of Environmental Policy." *Resource and Energy Economics*, 22: 233–259.

Pindyck, Robert S. 2007. "Uncertainty in Environmental Economics." *Review of Environmental Economics and Policy*, 1(1): 45–65.

Pindyck, Robert S. 2011a. "Fat Tails, Thin Tails, and Climate Change Policy." *Review of Environmental Economics and Policy*, 5(2): 258–274.

Pindyck, Robert S. 2011b. "Modeling the Impact of Warming in Climate Change Economics." In *The Economics of Climate Change: Adaptations Past and Present*, ed. G. Libecap and R. Steckel. University of Chicago Press.

Pindyck, Robert S. 2012. "Uncertain Outcomes and Climate Change Policy." *Journal of Environmental Economics and Management*, 63: 289–303.

Pindyck, Robert S. 2013a. "Climate Change Policy: What Do the Models Tell Us?" *Journal of Economic Literature*, 51(3): 860–872.

Pindyck, Robert S. 2013b. "The Climate Policy Dilemma." *Review of Environmental Economics and Policy*, 7(2): 219–237.

Pindyck, Robert S. 2013c. "Pricing Carbon When We Don't Know the Right Price." *Regulation*, 36(2): 43–46.

Pindyck, Robert S. 2014. "Risk and Return in the Design of Environmental Policy." *Journal of the Association of Environmental and Resource Economists*, 1(3): 395–418.

Pindyck, Robert S. 2017a. "Taxes, Targets, and the Social Cost of Carbon." *Economica*, 84(335): 345–364.

Pindyck, Robert S. 2017b. "The Use and Misuse of Models for Climate Policy." *Review of Environmental Economics and Policy*, 11(1): 100–114.

Pindyck, Robert S. 2019. "The Social Cost of Carbon Revisited." *Journal of Environmental Economics and Management*, 94: 140–160.

Pindyck, Robert S. 2021. "What We Know and Don't Know about Climate Change, and Implications for Policy." In *Environmental and Energy Policy and the Economy, Volume 2*, ed. M. Kotchen, J. Stock, and C. Wolfram. University of Chicago Press.

Pindyck, Robert S., and Daniel L. Rubinfeld. 2018. *Microeconomics, Ninth Edition.* New York: Pearson.

Pongratz, Julia, David B. Lobell, L. Cao, and Ken Caldeira. 2012. "Crop Yields in a Geoengineered Climate." *Nature Climate Change*, 2(2): 101–105.

Posner, Richard A. 2004. *Catastrophe: Risk and Response.* New York: Oxford University Press.

Rafaty, Ryan, Geoffroy Dolphin, and Felix Pretis. 2020. "Carbon Pricing and the Elasticity of CO_2 Emissions." Energy Policy Research Group, University of Cambridge Working Paper 2035.

Ramankutty, Navin, Holly K. Gibbs, Frédéric Achard, Ruth Defries, Jonathan A. Foley, and R. A. Houghton. 2007. "Challenges to Estimating Carbon Emissions from Tropical Deforestation." *Global Change Biology*, 13(1): 51–66.

Ranjan, Manya, and Howard J. Herzog. 2011. "Feasibility of Air Capture." *Energy Procedia*, 4: 2869–2876.

Reguero, Borja G., Michael W. Beck, David N. Bresch, Juliano Calil, and Imen Meliane. 2018. "Comparing the Cost Effectiveness of Nature-Based and Coastal Adaptation: A Case Study from the Gulf Coast of the United States." *PloS One*, 13(4).

Robock, Alan. 2000. "Volcanic Eruptions and Climate." *Reviews of Geophysics*, 38(2): 191–219.

Robock, Alan. 2020. "Benefits and Risks of Stratospheric Solar Radiation Management for Climate Intervention (Geoengineering)." *The Bridge*, 50: 59–67.

Robock, Alan, Allison Marquardt, Ben Kravitz, and Georgiy Stenchikov. 2009. "Benefits, Risks, and Costs of Stratospheric Geoengineering." *Geophysical Research Letters*, 36(19).

Roe, Gerard H., and Marcia B. Baker. 2007. "Why is Climate Sensitivity So Unpredictable?" *Science*, 318: 629–632.

Romm, Joseph. 2018. *Climate Change: What Everyone Needs to Know*. New York: Oxford University Press.

Rudik, Ivan. 2020. "Optimal Climate Policy When Damages are Unknown." *American Economic Journal: Economic Policy*, 12: 340–373.

Sanz-Pérez, Eloy S., Christopher R. Murdock, Stephanie A. Didas, and Christopher W. Jones. 2016. "Direct Capture of CO_2 from Ambient Air." *Chemical Reviews*, 116(19): 11840–11876.

Schaefer, Hinrich. 2019. "On the Causes and Consequences of Recent Trends in Atmospheric Methane." *Current Climate Change Reports*, 5: 259–274.

Schlenker, Wolfram, and Michael J. Roberts. 2009. "Nonlinear Temperature Effects Indicate Severe Damages to U.S. Crop Yields under Climate Change." *Proceedings of the National Academy of Sciences*, 106(37): 15594–15598.

Schuur, Edward A. G., A. David McGuire, C. Schädel, Guido Grosse, J. W. Harden, Daniel J. Hayes, Gustaf Hugelius, Charles D. Koven, Peter Kuhry, David M. Lawrence, et al. 2015. "Climate Change and the Permafrost Carbon Feedback." *Nature*, 520(7546): 171–179.

SDSN 2020. 2020. "Zero Carbon Action Plan." Sustainable Development Solutions Network Technical Report.

Sherwood, S., M. J. Webb, J. D. Annan, K. C. Armour, P. M. Forster, J. C. Hargreaves, G. Hegerl, S. A. Klein, K. D. Marvel, E. J. Rohling, M. Watanabe, T. Andrews, P. Braconnot, C. S. Bretherton, G. L. Foster, Z. Hausfather, A. S. von der Heydt, R. Knutti, T. Mauritsen, J. R. Norris, C. Proistosescu, M. Rugenstein, G. A. Schmidt, K. B. Tokarska, and M. D. Zelinka. 2020. "An Assessment of Earth's Climate Sensitivity Using Multiple Lines of Evidence." *Reviews of Geophysics*, 58(3).

Shindell, D. T., J. S. Fuglestvedt, and W. J. Collins. 2017. "The Social Cost of Methane: Theory and Applications." *Faraday Discussions*, 200: 429–451.

Skeie, Ragnhild Bieltvedt, Terje Koren Berntsen, Magne Tommy Aldrin, Marit Holden, and Gunnar Myhre. 2018. "Climate Sensitivity Estimates—Sensitivity to Radiative Forcing Time Series and Observational Data." *Earth System Dynamics*, 9(2): 879–894.

Smith, Jordan P., John A. Dykema, and David W. Keith. 2018. "Production of Sulfates Onboard an Aircraft: Implications for the Cost and Feasibility of Stratospheric Solar Geoengineering." *Earth and Space Science*, 5(4): 150–162.

Smith, Pete, Julio Friedmann, et al. 2017. "Bridging the Gap: Carbon Dioxide Removal." United Nations Environment Programme, The Emissions Gap Report 2017, Chapter 7.

Smith, Wake, and Gernot Wagner. 2018. "Stratospheric Aerosol Injection Tactics and Costs in the First 15 Years of Deployment." *Environmental Research Letters*, 13(12): 1–11.

Sokolov, Andrei, Sergey Paltsev, Henry Chen, Martin Haigh, Ronald Prinn, and Erwan Monier. 2017. "Climate Stabilization at 2°C and Net Zero Carbon Emissions." MIT Joint Program on Global Change Report 309.

Solomon, Susan, Gian-Kasper Plattner, Reto Knutti, and Pierre Friedlingstein. 2009. "Irreversible Climate Change due to Carbon Dioxide Emissions." *Proceedings of the National Academy of Sciences*, 106(6): 1704–1709.

Solomon, Susan, Martin Manning, Melinda Marquis, Dahe Qin, et al. 2007. *Climate Change 2007–The Physical Science Basis: Working Group I Contribution to the Fourth Assessment Report of the IPCC*. Vol. 4, Cambridge University Press: New York.

Sovacool, Benjamin K., Rasmus Andersen, Steven Sorensen, Kenneth Sorensen, Victor Tienda, Arturas Vainorius, Oliver Marc Schirach, and Frans Bjørn-Thygesen. 2016. "Balancing Safety with Sustainability: Assessing the Risk of Accidents for Modern Low-Carbon Energy Systems." *Journal of Cleaner Production*, 112: 3952–3965.

Stammer, Detlef, Anny Cazenave, Rui M. Ponte, and Mark E. Tamisiea. 2013. "Causes for Contemporary Regional Sea Level Changes." *Annual Review of Marine Science*, 5: 21–46.

Stavins, Robert. 2019. "The Future of U.S. Carbon-Pricing Policy." National Bureau of Economic Research Working Paper 25912.

Stern, Nicholas. 2013. "The Structure of Economic Modeling of the Potential Impacts of Climate Change Has Grafted Gross Underestimation onto Already Narrow Science Models." *Journal of Economic Literature*, 51(3): 838–859.

Stern, Nicholas. 2015. *Why Are We Waiting? The Logic, Urgency, and Promise of Tackling Climate Change*. Cambridge, MA: MIT Press.

Stocker, Thomas F., Dahe Qin, Gian-Kasper Plattner, Melinda Tignor, Simon K. Allen, Judith Boschung, Alexander Nauels, Yu Xia, Vincent Bex, Pauline M. Midgley, et al. 2013. "Climate Change 2013—The Physical Science Basis: Working Group I Contribution to the Fifth Assessment Report of the IPCC." *Intergovernmental Panel on Climate Change*, 1535.

Stock, James H. 2019. "Climate Change, Climate Policy, and Economic Growth." In *NBER Macroeconomics Annual*. , ed. Martin S. Eichenbaum, Erik Hurst, and Jonathan A. Parker. University of Chicago Press.

Temmerman, Stijn, Patrick Meire, Tjeerd J. Bouma, Peter M. J. Herman, Tom Ysebaert, and Huib J. De Vriend. 2013. "Ecosystem-Based Coastal Defence in the Face of Global Change." *Nature*, 504(7478): 79–83.

Ter Steege, Hans, Nigel C. A. Pitman, Daniel Sabatier, Christopher Baraloto, Rafael P. Salomão, Juan Ernesto Guevara, Oliver L. Phillips, Carolina V. Castilho, William E. Magnusson, Jean-François Molino, et al. 2013. "Hyperdominance in the Amazonian Tree Flora." *Science*, 342(6156).

Tilmes, Simone, Rolando R. Garcia, Douglas E. Kinnison, Andrew Gettelman, and Philip J. Rasch. 2009. "Impact of Geoengineered Aerosols on the Ttroposphere and Sstratosphere." *Journal of Geophysical Research: Atmospheres*, 114(D12).

Tjiputra, J. F., A. Grini, and H. Lee. 2016. "Impact of Idealized Future Stratospheric Aerosol Injection on the Large-Scale Ocean and Land Carbon Cycles." *Journal of Geophysical Research: Biogeosciences*, 121(1): 2–27.

Tol, Richard S. J. 2002a. "Estimates of the Damage Costs of Climate Change, Part I: Benchmark Estimates." *Environmental and Resource Economics*, 21: 47–73.

Tol, Richard S. J. 2002b. "Estimates of the Damage Costs of Climate Change, Part II: Dynamic Estimates." *Environmental and Resource Economics*, 21: 135–160.

Tol, Richard S. J. 2018. "The Economic Impacts of Climate Change." *Review of Environmental Economics and Policy*, 12(1): 4–25.

Ulph, Alistair, and David Ulph. 1997. "Global Warming, Irreversibility and Learning." *The Economic Journal*, 107(442): 636–650.

United Nations Environment Programme. 2020. "Emissions Gap Report 2020." United Nations Environment Programme.

U.S. Global Change Research Program. 2018. *The Climate Report: The National Climate Assessment—Impacts, Risks, and Adaptation in the United States*. Melville House: New York.

van Dantzig, David. 1956. "Economic Decision Problems for Flood Prevention." *Econometrica*, 24(3): 276–287.

van den Bremer, Ton S., and Frederick van der Ploeg. 2021. "The Risk-Adjusted Carbon Price." *American Economic Review*, 111(9): 2782–2810.

van Wesenbeeck, Bregje K., Wiebe de Boer, Siddharth Narayan, Wouter R. L. van der Star, and Mindert B. de Vries. 2017. "Coastal and Riverine Ecosystems as Adaptive Flood

Defenses under a Changing Climate." *Mitigation and Adaptation Strategies for Global Change*, 22(7): 1087–1094.

Vermeer, Martin, and Stefan Rahmstorf. 2009. "Global Sea Level Linked to Global Temperature." *Proceedings of the National Academy of Sciences*, 106(51): 21527–21532.

Vousdoukas, Michalis I., Lorenzo Mentaschi, Evangelos Voukouvalas, Martin Verlaan, and Luc Feyen. 2017. "Extreme Sea Levels on the Rise Along Europe's Coasts." *Earth's Future*, 5(3): 304–323.

Wagner, Gernot, and Martin L. Weitzman. 2015. *Climate Shock: The Economic Consequences of a Hotter Planet*. Princeton University Press.

Ward, Philip J., Brenden Jongman, Jeroen C. J. H. Aerts, Paul D. Bates, Wouter J. W. Botzen, Andres Diaz Loaiza, Stephane Hallegatte, Jarl M. Kind, Jaap Kwadijk, Paolo Scussolini, et al. 2017. "A Global Framework for Future Costs and Benefits of River-Flood Protection in Urban Areas." *Nature Climate Change*, 7(9): 642–646.

Ward, Philip J., Brenden Jongman, Peter Salamon, Alanna Simpson, Paul Bates, Tom De Groeve, Sanne Muis, Erin Coughlan De Perez, Roberto Rudari, Mark A. Trigg, et al. 2015. "Usefulness and Limitations of Global Flood Risk Models." *Nature Climate Change*, 5(8): 712–715.

Watson, James E. M., Tom Evans, Oscar Venter, Brooke Williams, Ayesha Tulloch, Claire Stewart, Ian Thompson, Justina C. Ray, Kris Murray, Alvaro Salazar, et al. 2018. "The Exceptional Value of Intact Forest Ecosystems." *Nature Ecology & Evolution*, 2(4): 599–610.

Weisenstein, D. K., D. W. Keith, and J. A. Dykema. 2015. "Solar Geoengineering Using Solid Aerosol in the Stratosphere." *Atmospheric Chemistry and Physics*, 15(20): 11835–11859.

Weitzman, Martin L. 2009. "On Modeling and Interpreting the Economics of Catastrophic Climate Change." *Review of Economics and Statistics*, 91: 1–19.

Weitzman, Martin L. 2011. "Fat-Tailed Uncertainty and the Economics of Climate Change." *Review of Environmental Economics and Policy*, 5(2): 275–292.

Weitzman, Martin L. 2014a. "Can Negotiating a Uniform Carbon Price Help to Internalize the Global Warming Externality?" *Journal of the Association of Environmental and Resource Economists*, 1(1): 29–49.

Weitzman, Martin L. 2014b. "Fat Tails and the Social Cost of Carbon." *American Economic Review*, 104(5): 544–546.

Weitzman, Martin L. 2015. "Internalizing the Climate Externality: Can a Uniform Price Commitment Help?" *Economics of Energy & Environmental Policy*, 4(2): 37–50.

Weitzman, Martin L. 2017. "On a World Climate Assembly and the Social Cost of Carbon." *Economica*, 84: 559–586.

Williamson, Phillip, and Carol Turley. 2012. "Ocean Acidification in a Geoengineering Context." *Philosophical Transactions of the Royal Society A*, 370: 4317–4342.

Zickfeld, Kirsten, and Tyler Herrington. 2015. "The Time Lag Between a Carbon Dioxide Emission and Maximum Warming Increases with the Size of the Emission." *Environmental Research Letters*, 10(3): 1–3.

Zickfeld, Kirsten, Michael Eby, Andrew J. Weaver, Kaitlin Alexander, Elisabeth Crespin, Neil R. Edwards, et al. 2013. "Long-Term Climate Change Commitment and Reversibility: An EMIC Intercomparison." *Journal of Climate*, 26(16): 5782–5809.

Index

Note: Tables are indicated by a *t*, Figures are indicated by a *f*.